Computational Design Modelling

Christoph Gengnagel, Axel Kilian, Norbert Palz,
and Fabian Scheurer (Eds.)

Computational Design Modelling

Proceedings of the Design Modelling
Symposium Berlin 2011

Editors

Prof. Dr. Christoph Gengnagel
Universität der Künste Berlin
Hardenbergstraße 33
10623 Berlin, Germany
E-mail: gengnagel@udk-berlin.de

Prof. Axel Kilian PhD
Princeton University
Princeton NJ 08544
USA

Prof. Dipl.-Ing. Norbert Palz
Universität der Künste Berlin UDK
Hardenbergstraße 33
10623 Berlin, Gemany
E-mail: n.palz@udk-berlin.de

Fabian Scheurer
designtoproduction GmbH
Seestraße 78
Erlenbach/Zurich, Switzerland

ISBN 978-3-642-23434-7   e-ISBN 978-3-642-23435-4

DOI 10.1007/978-3-642-23435-4

Library of Congress Control Number: 2011935739

© 2011 Springer-Verlag Berlin Heidelberg

This work is subject to copyright. All rights are reserved, whether the whole or part of the material is concerned, specifically the rights of translation, reprinting, reuse of illustrations, recitation, broadcasting, reproduction on microfilm or in any other way, and storage in data banks. Duplication of this publication or parts thereof is permitted only under the provisions of the German Copyright Law of September 9, 1965, in its current version, and permission for use must always be obtained from Springer. Violations are liable to prosecution under the German Copyright Law.

The use of general descriptive names, registered names, trademarks, etc. in this publication does not imply, even in the absence of a specific statement, that such names are exempt from the relevant protective laws and regulations and therefore free for general use.

Typeset & Cover Design: Scientific Publishing Services Pvt. Ltd., Chennai, India.

Printed on acid-free paper

9 8 7 6 5 4 3 2 1

springer.com

Foreword

Now in its third edition, the Design Modelling Symposium Berlin constitutes a platform for dialogue on experimental practice and research within the field of computationally informed architectural design.

Contemporary architectural production employs an increasing number of computational tools that undergo continuous proliferation of functions and expand their role within the design process. CAD/CAM technologies have matured into applications with increasingly user-friendly programme structures and an efficient exchange between various analytical tools. Computational geometry enables the design and manufacturing of complex surface configurations, a capacity beyond the repertoire of analog architectural practices constrained by the limitations of descriptive geometry. CAD/CAM technologies have been used successfully to achieve novel architectural expression by enabling digital geometry to drive digital fabrication processes. These innovations that have changed the work flow and design approach of a wide range of architectural practices and within academia.

Yet in parallel to these advances, limitations have become apparent. In many cases the relationship between design idea and computational tool seems reversed. The resulting buildings appear as reductionist materialization of the possibilities of software that shaped them. Only few examples exist where computational tools are used to develop design solutions for complex building programs within a moderate budget, yet driven by a rich conceptual approach that ventures beyond established theoretical paradigms of computational practice.

On the basis of these observations, a critical evaluation of the relationships between tool, conceptual model and final materialization appears necessary and valuable. The promise of an increased role of computational processes in the design of architecture lies in the manifold solutions that exceed human calculative capacities. A good example is the integration of Finite Element Method (FEM) based analysis procedures and generative form finding methods. However, these processes depend strongly on boundary conditions induced in the problem setup defined by the architect or engineer. Each optimization—be it structural or environmental—therefore can only produce a result within the realm of the abstracted (computational) model, and in no way represents a final solution for the real world or even an indication for changing the design conceptually. The complexity of the interconnected and often conflicting information required to shape a building—be it explicitly describable or not—remains a challenge for contemporary computational processes. A future architectural practice needs to cultivate a critical awareness of such limitations in order to develop successful future strategies.

The critical dialogue that we envision and encourage at the Design Modelling Symposium Berlin 2011 should be achieved by a collective contemplation of these current approaches and their entwined technological developments. We would like to promote discussion on future strategies for a reasonable and innovative implementation of digital potentials guided by both responsibility towards processes and the consequences they initiate. The fact that the discipline of architecture has in recent decades turned towards a scientific *modus operandi*—a process that leads to a communally orchestrated establishment of a rich, reflected and globally shared reference body in accordance with the protocols of science—should prove advantageous for a dialogue on the relationship between computational tool, concept and practice. This scientific turn in architecture has manifested itself in hundreds of papers, case studies, doctoral research and peer-reviewed publications. This research covers manifold fields and include—among other topics—design theory, digital fabrication, computational form finding, geometry and pedagogy. The constructive atmosphere of the last Design Modelling Symposium and comparable events has created a community characterized by openness, scientific rigor and curiosity. It is fair to assume that in the coming years a proliferation, specification and broader application of the investigated concepts and tools will take place in building practice, potentially altering the availability and distribution of these research findings.

This editorial preface is the result of a shared perspective on the core qualities that we consider necessary for a constructive investigation of the actual and future challenges of computational design and architectural practice. These qualities are centered on a practice of scientific verifiability, shared availability of knowledge and a continuous and constructive reflective monitoring of the manifold developments. We have therefore chosen to identify four areas that are specifically relevant to the field of computational design, fabrication and architectural practice. The brief statements that follow address the conceptual view on the thinking models that were introduced in the fields of architecture and engineering; portray their boundary conditions in regard to a realization on site; and give an outlook on future changes of functional and structural tectonics within building components.

Models of Design in Computation

Models are at the Core of Scientific Thinking

Design and even more so computational design relies heavily on abstraction and models of thought. The challenge of any abstraction is the invention of a construct that can stand for the actual phenomena with a good enough approximation to allow for making accurate predictions about the future solely based on the abstract model. This is the core of science and reasoning and so essential to our culture that it is hard to single it out. The formation of new models may start out with a mental model which is fluid and fluctuating, shaped by thoughts, dismissed and resurrected as needed. Defining a more stable, externalized and rigorous model requires substantially more effort. Translating it into a computational model requires an additional level of rigor as it can be operated independently of its creator and be reused essentially as a black box process without further scrutiny by an unaware user.

Merging Model Rigor and Design Process

Creating new models is difficult and hence the tendency is to work with existing models of thought. In computation this is even more likely due to the reusability of algorithms in the form of code. This path of least resistance has led to a limited set of computational models for design being used over and over again. Therefore a key motivation for holding an international conference on design modeling is to enable the survey and discussion of different approaches to conceptual models and the translation of ideas into novel computational models. A second motivation is to encourage the often substantial research investment to develop new and better fitting computational models for design. The notion of an overall model here is not limited to a 3D geometric data set but refers rather to the holistic, abstract representation of the overall design process including the role of humans in the process. Abstract models for the design process have concrete consequences in the design results. It is therefore not a question of design philosophy but, as we push for more interdisciplinary design work to take place, a question of how the underlying model defines the outcome. Therefore it is a core responsibility of the field to push forward with more integrated models of design, to test their capability to deal with real world complexity, and to evaluate their potential for improving design results.

From Analysis to Simulation

From Analysis of Known Problems to Simulation of New Scenarios

From a historic perspective the use of computers in structural engineering began much earlier and under different starting conditions than in the practice of architecture. Crucial for a more holistic deployment of the computer was the development of hardware in the 1980s, which allowed the computer to become an everyday tool for structural engineers already in the 1990s. Contrary to the developments in architecture computing in structural engineering was used as a tool for the analysis of structures. Only later its use as tool to speed up and rationalize design representations followed.

Exemptions were the design and execution of tensile constructs such as cable net and membrane roofs or the mostly in compression shell structures. In these areas lie the beginnings of the use of computers as a design tool for form finding in combination with structural analysis. The form finding of these load bearing systems is based on the search for a membrane geometry, which represents equilibrium of tensile forces in the surface given the geometric edge and support conditions, as well as possible external loads. This inverse question is based on the assumption of a prescribed state of tension in a yet unknown geometry, is independent of deformations and therefore does not require any material definition. The first form finding methods such as the force density method use the possible numerical simplification that follow form this definition. Today, the constantly increasing computational performance of the hardware and the continued improvement of FEM allow for the combination of form finding under consideration of the materiality in direct combination with structural analysis. Through these steps a change is taking place in structural design and construction development from analysis to simulation.

Computational Design as Experiment

Computational design becomes an experiment which investigates the structural behavior of increasingly complex systems. Most important are the possibilities of investigating the interplay between a system's elements with external forces. The first crucial step of a simulation is always the definition of the model. While for a long time the foundation of the design process was adapting the structure to be designed to known static based models, today the process begins with the definition of a model that is fine tuned to the task at hand. Modeling the problem requires a new creativity on the part of the engineer as well as knowledge that involves, besides the numeric basics, a substantial amount of craft. Here material and assembly knowledge are essential. The possibilities of increasingly complex simulations open up the question as to what extent we are capable of realizing the simulation outcomes in physical structures. Therefore a goal could be to use the new simulation possibilities of complex interdependencies to arrive at simple technical solutions that stand out for their multi-functional behavior. This New Low Tech design is characterized by the use of multifunctional, robust, material- and energy-efficient constructions, based on the use of high complexity computational experiments and a deep understanding of materials and jointing technology.

Computational Controlled Fabrication in Architecture

Architecture Is Built from Heterogeneous Components

The sheer size of buildings makes it practically impossible to fabricate a building as one homogeneous structure. There will always be building components that have to be assembled and connected in one way or another. In order to efficiently create large structures, the components have to reach a certain size, or the cost of assembly will become the main factor in the budget.

On top of that the multitude of different functions a building has to fulfill requires a multitude of different building materials. Since they all have their specific properties and most likely different fabrication technologies, designing and fabricating building components thus requires specialist knowledge from a multitude of domains—usually not found in one single place or brain. The integration of various functions into polyvalent components may reduce the number of different component types but increases the complexity and the embedded knowledge of each type while at the same time eliminating clear interfaces between different trades. Thus the integration of all required know-how through close cooperation of all involved parties becomes indispensible, also for handling the responsibility and risks of the process.

Digital Fabrication Means Pre-Fabrication

Apart from very few exceptions (e.g. robots for the rather monofunctional task of brick laying or the robotic high-rise-building "factories" that never really made it outside of Japan), digital fabrication equipment is too large, costly and delicate to move it to site and to build components directly at their final location. Thus, digital fabrication almost always means pre-fabrication of building components in the controlled environment of the fabricator's workshop, shipping them to the site and assembling them like a big puzzle. That adds a number of challenges, mainly for just-in-time procurement, production and logistics that have to be carefully dealt with before the actual building process can start. The planning effort has to be moved almost completely to the front of the process, since finding mistakes during the assembly on-site and far away from the fabrication facilities can become catastrophic in terms of budget and design intent.

Digital Fabrication Needs Precise Descriptions

Computers are deterministic machines and need correct input to deliver correct output. That also holds true for the controllers of CNC-fabrication machines. Every drilling, milling, bending, planning, glueing or cutting operation a machine has to execute must be unambiguously defined in the digital model that is fed into the machine, down to last screw hole. In general it is not sufficient to create a 3D-model of the component to be produced, because aside from simple 2D-cutting or 3D-printing operations, the translation of geometric description into the machining sequence of a multi-axis CNC-machine, maybe involving several tool-changes during the process, is far from a linear problem. So, to come up with a working fabrication model for a complex component requires the full set of production knowledge for the specific machinery used.

Digital Fabrication Is First and Foremost a Question of the Process

The technology for digital fabrication is widely available today. The challenge now is to understand how those machines can be integrated into the existing processes of building and how this might change the traditional way to design in order to exploit the full capabilities of the fabrication equipment. On the other hand, through a deeper understanding of the current technology, we will be able to better identify shortcomings and specific needs of the building sector and (re-) direct the development of future technologies to better fit the requirements of the architectural process.

Rapid Manufacturing in Architecture

Fabrication of Material and Structural Heterogeneity

Rapid manufacturing has proliferated in the past years due to an improvement of additive fabrication (AF) processes in regard to mechanical properties, greater material diversity and scale of the producible artifacts. In architecture AF can exceed prior representative applications and progress towards the fabrication of functioning components of high complexity of structure and material composition. First signs indicate such a potential implementation through research conducted by several private and academic institutions on additively fabricated building-scale parts. Technological progress is accompanied by recent standardization efforts of AF processes and material quality through academic institutions and industry that can promote utilization of AF components within the building sector.

Yet additive fabrication of functional building parts requires a phase of wider experimental investigation to be conducted in the coming years. An interesting segment of contemporary AF research is hereby not only investigating the production possibilities of specialized parts but also the calibration of the material itself with regard to its structural performance and composition control. The benefit of this research lies in an achievable congruence between technological development and design activity once the AF processes are suited for architectural applications.

The envisioned digitally driven calibration and construction of novel structures and formations hereby alters the historical dialogue on material, structural and formal coherency. The conceptual approach between construction typology and material use that persisted architectural history until now is about to change again. The question in the future will not be centered on a best fitting structural solution for a given material with more or less known properties that drove the thinking of Viollet-le-Duc and others, but a reverse process that tailors a custom material with gradual and non-repetitive characteristics to a chosen form and performance.

Towards a New Building Tectonic

The achievable control over structure, material, and form opens up a design potential that is a direct descendant of the core properties of the fabrication tectonic and can give birth to novel building components. On a constructive level a merging of multiple building functions into a singular component appears achievable. The timeline of assembly that is usually coordinated from the erecting of a primary load-bearing structure downwards and shapes the appearance of the buildings around us can potentially blend multiple building functions in a new construction component whose dimensions are then based on building chamber sizes of the manufacturing technology. Formal complexity and ease of assembly through new joinery systems can so be achieved. The designed structural morphology could be guided by shape- and topology optimization procedures and by that integrate a material saving building practice in the load-bearing core of the project. This rethinking of architectural, structural and material practice holds many promises and a manifold of technological challenges that will take decades to be overcome.

The impression that a wide range of functionalities can be tuned and optimized might be misleading in context of potentially opposing optimization goals that have to be synchronized. Beside such restricting aspects research in these areas is of great interest since it holds a manifold of possible innovations for the design and construction process within architecture and may lead to the rewriting of the historical discussion on the relation between matter and form.

The following collection of papers presented at the Design Modelling Symposium Berlin 2011 is a cross section through current cutting edge research in the field. Some of the papers respond to the challenges and questions formulated above, others open up new discourses departing from the topics outlined here. Most importantly, all authors succeed in challenging our current understanding of the field through the rigor of the presented work. In doing so, they foster advances in architecture and engineering as well as the discourse that creates the conceptual basis of our disciplines.

Christoph Gengnagel, University of the Arts, Berlin
Axel Kilian, Princeton University, Princeton
Norbert Palz, University of the Arts, Berlin
Fabian Scheurer, designtoproduction, Zurich

Contents

Foreword V
Christoph Gengnagel, Axel Kilian, Norbert Palz, Fabian Scheurer

Concept, Tool and Design Strategies

DesignScript: Origins, Explanation, Illustration 1
Robert Aish

Algebraic Expansions: Broadening the Scope of Architectural Design through Algebraic Surfaces 9
Günter Barczik, Daniel Lordick, Oliver Labs

Tools and Design Strategies to Study Rib Growth 17
Chris Bardt, Michal Dziedziniewicz, Joy Ko

Free Shape Optimal Design of Structures 25
Kai-Uwe Bletzinger

NetworkedDesign, Next Generation Infrastructure for Design Modelling 39
Jeroen Coenders

Digital Technologies for Evolutionary Construction 47
Jan Knippers

Combinatorial Architecture 55
Enrique Sobejano

Codes in the Clouds Observing New Design Strategies 63
Liss C. Werner

Modeling, Simulation and Optimization

Methodological Research

Integration of Behaviour-Based Computational and Physical Models: Design Computation and Materialisation of Morphologically Complex Tension-Active Systems 71
Sean Ahlquist, Achim Menges

Synthetic Images on Real Surfaces 79
Marc Alexa

Modelling Hyperboloid Sound Scattering: The Challenge of Simulating, Fabricating and Measuring 89
Jane Burry, Daniel Davis, Brady Peters, Phil Ayres, John Klein, Alexander Pena de Leon, Mark Burry

Integration of FEM, NURBS and Genetic Algorithms in Free-Form Grid Shell Design 97
Milos Dimcic, Jan Knippers

SOFT.SPACE_Analog and Digital Approaches to Membrane Architecture on the Example of Corner Solutions 105
Günther H. Filz

Performance Based Interactive Analysis 115
Odysseas Georgiou

On the Materiality and Structural Behaviour of Highly-Elastic Gridshell Structures 123
Elisa Lafuente Hernández, Christoph Gengnagel, Stefan Sechelmann, Thilo Rörig

Parametric Design and Construction Optimization of a Freeform Roof Structure 137
Johan Kure, Thiru Manickam, Kemo Usto, Kenn Clausen, Duoli Chen, Alberto Pugnale

Curved Bridge Design 145
Lorenz Lachauer, Toni Kotnik

Linear Folded (Parallel) Stripe(s) 153
Rupert Maleczek

The Potential of Scripting Interfaces for Form and Performance Systemic Co-design 161
Julien Nembrini, Steffen Samberger, André Sternitzke, Guillaume Labelle

Contents

Building and Plant Simulation Strategies for the Design of Energy Efficient Districts 171
Christoph Nytsch-Geusen, Jörg Huber, Manuel Ljubijankic

New Design and Fabrication Methods for Freeform Stone Vaults Based on Ruled Surfaces 181
Matthias Rippmann, Philippe Block

Design and Optimization of Orthogonally Intersecting Planar Surfaces 191
Yuliy Schwartzburg, Mark Pauly

Modelling the Invisible 201
Achim Benjamin Späth

Applied Research

Ornate Screens – Digital Fabrication 209
Daniel Baerlecken, Judith Reitz, Arne Künstler, Martin Manegold

The Railway Station "Stuttgart 21": Structural Modelling and Fabrication of Double Curved Concrete Surfaces 217
Lucio Blandini, Albert Schuster, Werner Sobek

Performative Surfaces: Computational Form Finding Processes for the Inclusion of Detail in the Surface Condition 225
Matias del Campo, Sandra Manninger

ICD/ITKE Research Pavilion: A Case Study of Multi-disciplinary Collaborative Computational Design 239
Moritz Fleischmann, Achim Menges

Metropol Parasol - Digital Timber Design 249
Jan-Peter Koppitz, Gregory Quinn, Volker Schmid, Anja Thurik

Performative Architectural Morphology: Finger-Joined Plate Structures Integrating Robotic Manufacturing, Biological Principles and Location-Specific Requirements 259
Oliver Krieg, Karola Dierichs, Steffen Reichert, Tobias Schwinn, Achim Menges

A Technique for the Conditional Detailing of Grid-Shell Structures: Using Cellular Automata's as Decision Making Engines in Large Parametric Model Assemblies 267
Alexander Peña de Leon, Dennis Shelden

Parameterization and Welding of a Knotbox 275
Daniel Lordick

Viscous Affiliation - A Concrete Structure 283
Martin Oberascher, Alexander Matl, Christoph Brandstätter

Dynamic Double Curvature Mould System 291
Christian Raun, Mathias K. Kristensen, Poul Henning Kirkegaard

More Is Arbitrary: Music Pavilion for the Salzburg 301
Kristina Schinegger, Stefan Rutzinger

Design Environments for Material Performance 309
*Martin Tamke, Mark Burry, Phil Ayres, Jane Burry,
Mette Ramsgaard Thomsen*

Educational Projects

Faserstrom Pavilion: Charm of the Suboptimal 319
Mathis Baumann, Clemens Klein, Thomas Pearce, Leo Stuckardt

**Rhizome - Parametric Design Inspired by Root Based Linking
Structures**... 327
Reiner Beelitz, Julius Blencke, Stefan Liczkowski, Andreas Woyke

Kinetic Pavilion: Extendible and Adaptable Architecture 335
Corneel Cannaerts

Author Index 341

DesignScript: Origins, Explanation, Illustration

Robert Aish

> "A programming language that doesn't change the way you think is not worth learning"
>
> —Alan Perlis, 'Epigrams in Programming'

Abstract. DesignScript, as the name suggests, is positioned at the intersection of design and programming. DesignScript can be viewed as part of the continuing tradition of the development of parametric and associative modeling tools for advanced architectural design and building engineering. Much of the thought processes that contribute to the effective use of DesignScript builds on the tradition of parametric design and associative modeling that is already widely distributed amongst the creative members of the architectural and engineering communities. Many of the existing parametric and associative modelling tools also support conventional scripting via connections to existing programming languages. The originality of DesignScript is that associative and parametric modeling is integrated with conventional scripting. Indeed, the definition of the associative and parametric model is recorded directly in DesignScript. But it is not what DesignScript *does* which is important, more what a designer can *do* with DesignScript. It is this change in the way you think that makes DesignScript worth learning.

1 Introduction

DesignScript is intended to be:

- a production modeling tool: to provide an efficient way for pragmatic designers to generate and evaluate complex geometric design models
- a fully-fledged programming language: as expected by expert programmers.
- a pedagogic tool: to help pragmatic design professions make the transition to competent programmer by the progressive acquisition of programming concepts and practice applied to design.

Robert Aish
Director of Software Development, Autodesk

Essentially there are three themes interwoven here:

- The programming language theme: DesignScript as a programming language
- The design process theme: The use of DesignScript as a design toolset
- The pedagogic theme: using DesignScript as a way of learning how to design and to program.

2 Programming Language

From the perspective of a programming language, we might describe DesignScript as an *associative* language, which maintains a graph of dependencies between variables. In DesignScript these variables can represent numeric values or geometric entities, or other application constructs, including those defined by the user. The execution of a DesignScript program is effectively a *change-propagation* mechanism using this graph of variables. This change-propagation also functions as the update mechanism similar to that found in a conventional CAD application. However, unlike other CAD update mechanisms or associative and parametric modeling systems, in DesignScript this mechanism is exposed to the user and is completely programmable. Figure 1 illustrates the important differences between a conventional *imperative* language and an *associative* language such as DesignScript, while Figure 2 shows how a program statement in DesignScript can also be interpreted as natural language. Each term in the statement has an equivalent natural language interpretation so that whole statement can be understood by its natural language equivalent.

So a concise but somewhat complex description of DesignScript might be as a domain-specific, end-user, multi-paradigm, host-independent, extensible programming language (Fig. 3), as follows:

1. **Domain-specific.** DesignScript is intended to support the generation of geometric design models and therefore provides special constructs to assist in the representation of geometric models. More generally: A domain specific language may remove certain general purpose functionality and instead adds domain specific functionality as first class features of the language.
2. **End-user.** DesignScript is intended to be used by experienced designers with a wide range of programming skill, ranging from non-programmers (who might indirectly *program* via interactive direct manipulation), to novice non-professional (end-user) programmers, and to experienced designers who have substantial expertise in programing. More generally: An end user language adds simplifying syntax to the language, while reducing some of restriction often associated with general purpose languages (intended for experienced programmers).
3. **Multi-paradigm.** DesignScript integrates a number of different programming paradigms into a single language (including object-oriented, functional and associative paradigms) and introduces some additional programming concepts

DesignScript: Origins, Explanation, Illustration

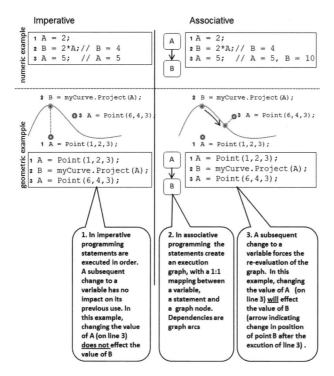

Fig. 1 Comparing Imperative and Associative interpretation of the same program statements. It is this *change in the way you think* that makes DesignScript worth learning.

Fig. 2 Giving a natural language interpretation to a DesignScript statement

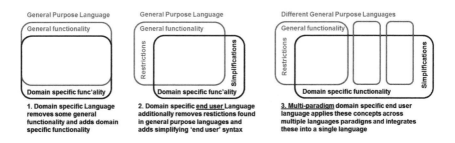

Fig. 3 How DesignScript differs from a regular general purpose programming language

that are relevant to the domain of generative design. More generally: A multi-paradigm language combines different programming styles into a single language and allows the user to select which paradigms or combination of paradigms are appropriate. (See Fig. 4)
4. **Host-independent.** DesignScript is intended to support the generation of geometric models and is therefore designed to be hosted within different CAD applications and access different geometric, engineering analysis and simulation libraries. For example, a DesignScript variable (based on specific class) may maintain a correspondence with a geometric entity in AutoCAD and simultaneously with entities within engineering analysis applications such as Ecotect and Robot.
5. **Extensible.** DesignScript can be extended by the user, by the addition of functions and classes.

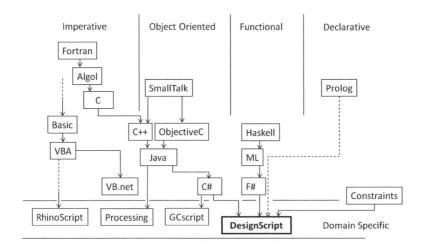

Fig. 4 The evolutionary *tree* for DesignScript (showing its precursors). DesignScript is a multi-paradigm language embracing imperative, objected oriented, functional and declarative programming concepts.

3 Design Process

DesignScript is intended to support a computational approach to design which is accessible to designers who initially may be unfamiliar with this way of designing. Conventionally, computer-based design applications enabled the designer to create models which represent finished designs. The intention in developing DesignScript is to move beyond the representation of finished designs, and instead to support the designer to develop his own geometric and logical framework within which many different alternative design solutions can be easily generated and evaluated.

The development of DesignScript assumes that the designer wants to adopt this more exploratory approach to design and that he appreciates that this may involve some re-factoring of the design process so as to include a more explicit externalization of particular aspects of design thinking, for example:

- Explicitly identifying the key variables that drive the design.
- Building the geometric and logical dependencies between these driver variables and the constructive geometry: potentially these dependencies can be complex *long chains*.
- Defining appropriate performance measures that can describe the resulting design solutions.
- Exercising the complete model (by changing the design drivers and observing changes in the geometry and resulting performance measures) to explore more appropriate solutions.
- Changing the geometric and logical dependencies in order to explore more alternatives.

4 Pedagogic Perspective

From a pedagogic perspective, DesignScript is designed around the concept of a learning curve and supports a very gradual approach to learning programming (Fig. 5):

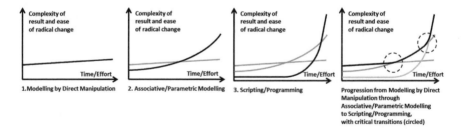

Fig. 5 DesignScript as conceived as a composite learning curve spanning different types of modelling and programming

1. **For modelling by *direct manipulation***, the designer immediately obtains some interesting result for the modelling effort he makes, yet to change or refine or increase the complexity of the model may require an exhaustive amount of additional effort. Therefore the perceptive designer may search for a way to overcome the limitations of direct manipulation.

2. **For Associative or parametric modelling**, the designer may have to initially make some more effort to create the first associative model (than he did with regular modelling). Although the initial results may be unimpressive, he is investing in an associative model with higher semantic value. Because of this investment in design logic the designers ability to change and refine that model becomes comparatively easy (compared to non-associative modelling). The designer is not just investing his time and effort, but also has to learn new skills: in particular how to think associatively. However, the perceptive designer may recognise that some types of design logic are difficult to express in an associative modelling system, therefore the perceptive designer may search for a way to overcome the limitations of associative modelling.
3. **With scripting and programming**, considerable time and effort may be expended apparently without much evidence of success. Nothing works until it all works, but then the complexity of the model and the ability to re-generate the model with radically different design logic appears more powerful than what can be achieved with associative modelling.

We can summarise this as:

- *Learning by doing*, for example, by interactive modelling
- *Learning by observing the correspondence between the DesignScript notation and geometry*, for example, by comparing the geometric model with the graph based symbolic model and with the DesignScript notation displayed in the IDE)

The following example illustrates the use of DesignScript. The design problem is to model a wave roof, based on a complex wave formation. The first step is to recognise that we should not attempt to directly model the wave formations with regular modelling tools. Instead we should recall that most complex wave forms can be constructed as the aggregate effect of simpler waves combined with related harmonic waves. In this case, the geometry is constructed by using a series of low and high frequency sine waves running orthogonally in the X and Y direction (Fig 6). The amplitude and number of peaks in the waves are controlled by root parameters. The X, Y and Z coordinates of the 2D field of points is defined by combining these sine waves (Fig 7). The number of peaks can be varied (Fig 8). The X, Y and base Z coordinates of the points can be derived from points in the UV parametric space of a surface, thereby giving the effect that the wave geometry is draped (and offset) from an underlying surface (Fig 9). Finally, the control vertices of the underlying surface can be modified giving the effect that the underlying surface is controlling the wave roof (Fig 10).

This presents the exactly the combination of direct modelling, associative modelling and scripting suggested in the learning curve in Fig. 5. It is not just the model (or the computation of the model) which is spanning this different approaches. It is the thought processes of the designer which is combining these different ways of thinking.

Fig. 6 High and Low frequency waves in the X and Y directions

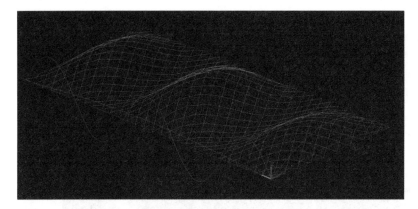

Fig. 7 The resulting wave roof is created by aggregating these orthogonal waves

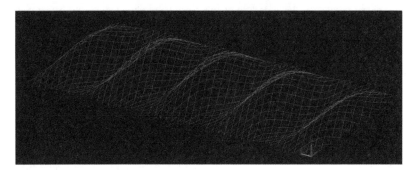

Fig. 8 The number of peaks can be varied

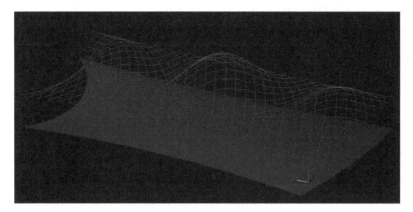

Fig. 9 Draping (and offsetting) the wave roof from an underlying surface

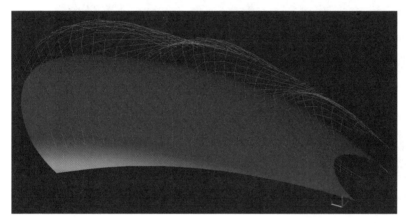

Fig. 10 The control vertices of the underlying surface can be modified giving the effect that the underlying surface is controlling the *wave* roof

5 Discussion

The three themes which are interwoven here (the programming language theme, the design process theme and the pedagogic theme) all come together when we address the central issue: How can a computational tools invoke a computational mindset and in turn contribute to design thinking?

Using DesignScript is a new way of designing with its own expressive possibilities. But there is a level of understanding required to harness this expressiveness and this suggests a level of rigor and discipline. The argument is that the experience of learning and using DesignScript contributes not just to the expressiveness and clarity of the resulting design but also to the skills and knowledge of the user.

In short, "a new toolset suggests a new mindset".

Algebraic Expansions: Broadening the Scope of Architectural Design through Algebraic Surfaces

Günter Barczik, Daniel Lordick, and Oliver Labs

1 Introduction: An Expanded Architectural Design Vocabulary

We conduct a design research project that radicalizes the relationship between tools and design possibilities: we significantly expand the architectural design vocabulary by employing mathematics and computer science as vehicles for accessing shapes that otherwise would be unthinkable: algebraic surfaces, the zero-sets of certain polynomials.

Algebraic surfaces can exhibit geometric features that cannot - or have so far not - be found in nature: puzzling convolutions in which complex geometry and topology combine with high degrees of tautness, harmony and coherence (Fig.1). Albeit mostly curved, they can contain straight lines and any number of plane curves (Fig.1, 1-3). They also look different from every direction, a quality we propose to call polyoptical from the Greek for an object with many faces. Having been studied in mathematics for the last two centuries they became accessible for designers only recently via advances in computer technology. This means a cambrian explosion of shapes, a whole zoo of new exotic shapes.

Günter Barczik
Brandenburg Technical University Cottbus, Germany
HMGB architects, Berlin, Germany

Daniel Lordick
Institute of Geometry, University of Technology Dresden, Germany

Oliver Labs
Mathematics and its Didactics, Cologne University, Germany
Institute for Mathematics and Computer Science, Saarbrücken University, Germany

Fig. 1 Examples of Algebraic Surfaces by Oliver Labs (1-4), Herwig Hauser (5,6) and Eduard Baumann (7,8)

2 Two Ways of Dealing with the Zoo of New Shapes

The new shapes can be dealt with in two ways: they can be taken literally, or as inspirational objects akin to Le Corbusiers Objects à rèaction poetique.

Corbusier collected pieces of wood corroded by water and wind and sea-shells to provoke thinking about geometrical and textural qualities. Over the years, this altered his designs from white boxes to buildings like the chapel in Ronchamp. In a similar way, algebraic surfaces can be employed to stimulate thinking about spatial configurations and relationships. Thus, they can foster a new understanding of the already existing plastic vocabulary.

Algebraic surfaces might also be taken literally and interpreted as buildings or parts of buildings. They then add many new words to the textbook of possible architectural shapes. In language, a large vocabulary enables speakers to phrase thoughts more precisely. Similarly, a large vocabulary of shapes should enable designers to formulate more appropriate solutions. We research this expansion in a series of experimental designs - see below.

3 Historic Precedents for Mathematics Inspiring Art and Architecture

Such use, or mis-use, of mathematical entities for design may appear contrived, yet it is neither without precedent nor without profound effect on the history of modern art and architecture: when in the first half of the 20th century artists like Naum Gabo or Man Ray formulated a constructivist agenda to add new things to the world, things that could not be generated through observation in or abstraction from nature, nor via surrealist drug-induced or aleatoric procedures, they were shocked to discover mathematical models which demonstrated that their goals had already been achieved a century earlier. Those mathematical models had been built to visualize algebraic surfaces and other mathematical objects. The Constructivists started to copy these mathematical objects in painting and sculpture, but they found themselves unable to understand them or generate their own ones as the calculations necessary for their generation could only be done by experts. Similarly in architecture, Le Corbusier handed Yannis Xenkis illustrations of mathematical functions as starting points for the design of the Philips Pavillon for the 1958 World Expo, taken from a book that had been sent to him from the faculty of mathematics of the University of Zrich after Corbusier had explicitly asked its dean for inspirational material. Comparable to the constructivist artists, Xenakis struggled with handling the new shapes. Due to such technical obstacles, those mid-century artists and architects mistook for a dead-end the road to great discoveries.

In recent years, sculptors like Anthony Cragg and Anish Kapoor have again begun to explicitly add new shapes to the world. Although they do not mention mathematics, the geometric possibilities attained through computers and shapes obviously related to mathematical objects feature heavily in their work.

Yet, the shapes employed by the mid-20th-century artists as well as those used by Cragg and Kapoor fall significantly short of the ones which can be generated via algebraic geometry in terms of complexity.

4 Five-Step Design Research Program

Our experimental design program is divided into five steps: generation, interpretation, adaptation, application and production.

First we generate the surfaces via the software packages Surfer, SingSurf and K3DSurf. All three accept a polynomial as input and output visualizations or/and 3D models. Surfer is restricted to visualization, but highly interactive. SingSurf and K3DSurf are not as interactive but generate 3D data that can be exported as polygon meshes. All programs do not determine the zero-sets of the polynomials by solving those equations exactly - currently no applicable software for this exists. Instead, they offer approximations, leading to inaccuracies in the models which occasionally show up as imperfections. Furthermore, the normals of the meshes are most usually disoriented and have to be aligned. Albeit such technical difficulties, in most cases the shapes that are the zero-sets can be successfully imported into CAD software. Once imported, they could as a matter of course be mimiced or re-built as a NURBS surface. The definition of the NURBS surface, though, would have to be based on points or polylines extracted from the mesh. So far, we have opted for smoothing out the meshes via the Catmull-Clark subdivision surface algorithm with most satisfactory results (Fig.2).

Secondly the surfaces are analyzed in terms of their geometric properties and interpreted as to their architectonic potential. The shapes exhibit exotic sculptural

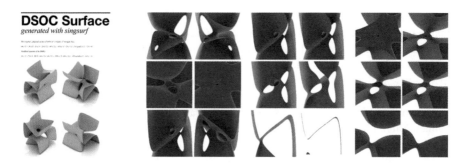

Fig. 2 Creating and analyzing an Algebraic Surface in perspective renderings and sections (Stefan Schreck)

situations that so far are unnamed: connections between different regions that are neither holes nor tunnels and might be named passages, self-intersections, singular points that mathematicians call singularities, to name but a few. The surfaces are mostly continuously curved and rarely flat. Therefore, they do not seem to invite architectonic use at first glance. Yet, as humans happily exploit non-flatness i.e. in undulating parkscapes where people sit, lie, play, gather and disperse in relation to the topography we see this more as an inspiration to question the prevalence of flatness that pervades modern architecture. There also is a strand within the avant-garde architecture of the last few decades that explicitly researches the use of non-flat surfaces, beginning with Claude Parents theory of the oblique and ending, so far, in Kazuyo Sejimas and Ryue Nishizawas Rolex Learning Centre in Lausanne and Sou Fujimotos Primitive Future House project. We pick up this strand to see if the flatland of modern architecture might not be expanded to more spatial configurations and more formfitting uses (Fig.3).

In a third step, the algebraic surfaces are adapted geometrically to facilitate humans use - i.e. stretched, twisted, compressed. Additionally, they are converted from

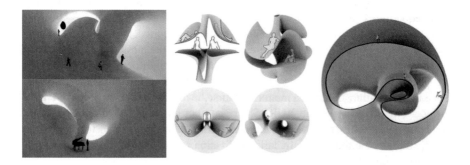

Fig. 3 Interpreting three algebraic surfaces as to their spatial potential (David Schwarzkopf, Dana Kummerlöw, Susann Seifert (from left to right))

Fig. 4 Adapting two algebraic surfaces into enclosed volumes (Dana Kummerlöw (left) and Christopher Jarchow (right))

surfaces into enclosures through various operations like for example section with a cuboid or deforming the surface until it becomes a volume (Fig.4).

In a fourth step, experimental architectures are generated by synthesizing the knowledge and know-how acquired in the first three steps (Fig.5-7). The resulting building designs are furthermore situated in urban contexts. While it can be argued that the extra-ordinary shapes of algebraic surfaces by definition have difficulty becoming part of any urban context, we argue that human settlements have always contained special buildings that have often been the most radical expression of what was possible at any given time. Those special buildings have also played important roles in the social life of communities, attracting visitors and inspiring social and cultural exchange. The building designs which incorporate the unprecedented shapes of algebraic surfaces can therefore play important roles in human communities, providing spatial focus points and inspiring new forms of social exchange. Additionally, the polyoptical qualities of such shapes [see above] means that they can relate differently to the more and more diversified urban fabrics of todays cities.

The last step consists of printing the designs in 3D (Fig.8). While we use the technology to print only models of the designs, it is rapidly progressing to print larger and larger objects, the largest at the moment exceeding telephone box size. While it is as a matter of course not satisfactory to see building construction as a matter of simply printing large objects of a uniform material, for us in our project the printability of the shapes is proof of concept enough insofar that unprecedented, new and never seen or touched objects are transported from the intellectual world of mathematics into a tangible physical reality.

5 Function Inspired by Form?

The steps we take in the experimental design project changes the common design procedure of Form follows function to Form inspires function or even Function follows form. At first, this may be seen as a severe restriction of designers capabilities, restraining their options to a corset defined by a given algebraic surface. Yet, we

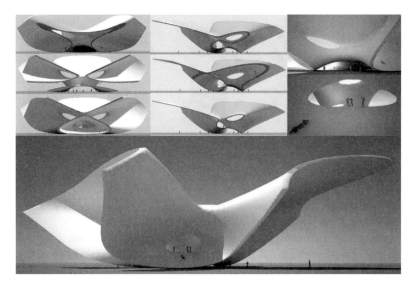

Fig. 5 Experimental design project based on an algebraic surface (Xing Jiang)

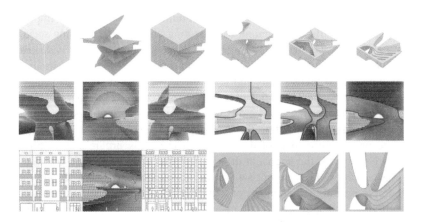

Fig. 6 Experimental design project based on an algebraic surface (Jörg Burkart)

understand our project merely as acquiring a new vocabulary. And in any such undertaking, existing new vocables have to be learned, played and experimented with before they can become part of the active vocabulary and used at will and as different situations and problems of formulation necessitate. This can also be seen in the way that children learn and get to know new shapes: nobody is born with a knowledge of eucledian geometry or, for that matter, any shape at all. Those have to be encountered in the world through perception and thus build up a spatial vocabulary. We argue that only when one forgets these learning experiences our procedure, mimicing them, appears wrong.

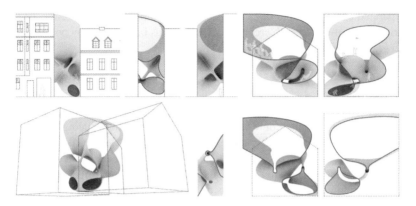

Fig. 7 Experimental design project based on an algebraic surface (Dana Kummerlöw)

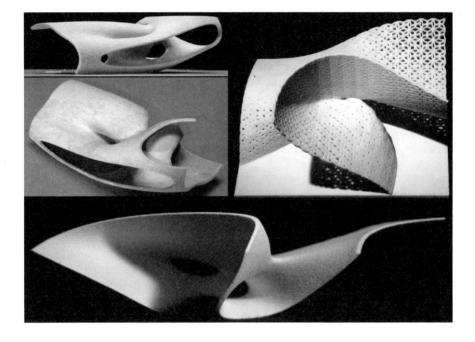

Fig. 8 Model prints of experimental design projects (Joanna Kollat (top left), Stefan Schreck (top right) and Xing Jiang (bottom))

6 Gradient Thresholds

Many algebraic surfaces clearly exhibit different regions of space with different geometric qualities. These regions are almost never exactly demarkated but flow gradually into one another. The threshold between them is not a line but a gradient. This can lead to a new kind of multifunctionality or hybrid use where the different zones are not seperated as i.e. different floor levels but share common areas of ambivalent use. The rigid territories of much architecture might thus be enriched by polyvalent areas with gradient thresholds.

7 Conclusion

Our project extends the architectonic vocabulary of shapes by introducing unprecedented new forms that until recently could not be thought let alone visualized or handled. This zoo of new shapes expands the possibilities of use of space, habitation and social interaction and offers alternatives to the flatland and rigid territories of most current architecture. Yet the process of getting to know, let alone mastering the new vocabulary has only just started, and there is indeed the danger of stopping here already and only revelling in appealing new shapes that are rather detached from architectonic design that integrates issues of organization, structure, context and so forth. We think, though, that learning a new vocabulary takes time, patience and much experimentation - especially when the language is completely new to thought and was never spoken before. Algebraic shapes made visible and useable through computers, we think, can continue several strands of research into architectural possibilities that have begun in the last century and reinvigorate them with unrecedented possibilities.

Acknowledgements. Thanks to all students and to the 3D Labs of the Universities of Dresden and Poznan where most of our models were printed.

References

1. Barczik, G., Labs, O., Lordick, D.: Algebraic Geometry in Architectural Design. In: Proceedings of the 27th eCAADe, Istanbul, Turkey (2009)
2. Barczik, G., Labs, O., Lordick, D.: Perplexing Beauty: The Aesthetics of Algebraic Geometry in Architecture. In: Proceedings of the IAEA 2010, Dresden (2010)
3. Barczik, G.: Uneasy Coincidence? Massive Urbanization and New Exotic Geometries with Algebraic Geometry as an extreme example. In: Proceedings of the 28th eCAADe, Zürich, Switzerland (2010)
4. Barczik, G.: Leaving Flatland behind. In: Proceedings of the 29th eCAADe, Ljubljana, Slovenia (forthcoming, 2011)
5. Maak, N.: Der Architekt am Strand, München (2010)
6. Cecilia, M., Levene.: El Croquis #155 Sou Fujimoto, Madrid (2011)
7. Migayrou, R.: Claude Parent: L'oeuvre construite, l'oeuvre graphique, Paris (2010)
8. Eduard Baumann's Algebraic Surfaces, http://www.spektrum.de/sixcms/list.php?page=p_sdwv_mathekunst&_z=798888&sv%5Bvt%5D=eduard+baumann&kategorie=%21Video&x=0&y=0
9. Herwig Hauser's Algebraic Surfaces, http://www.freigeist.cc/gallery.html

Tools and Design Strategies to Study Rib Growth

Chris Bardt, Michal Dziedziniewicz, and Joy Ko

1 Introduction

Ever since Viollet-le-Duc the 19th century engineer and architect proposed "natural" structures borne out of ideal forms of specific materials, engineers and architects alike have been interested in the notion of an organic approach to form and structure [7]. Functionalism, the idea that form is a resultant of forces and needs, was core to the modernist project but became overly deterministic and untenable for architecture [1]. The course of much of the latter part of the 20th century history was one of the separation of engineering (calculation) and architecture (organization) into two exclusive realms uneasily brought together, with one or the other taking the lead in the generation of form [4]. In the architectural design process, calculation of structural performance customarily entered late in the design process when the form was already largely realized.

Computation, now the widely accepted *lingua franca* of many fields that architecture touches, has played a central role in rekindling interest amongst architects to rejoin calculation and organization as a critical step in creating truly performative forms. Current interests in architecture such as biomimicry, genetic algorithms, and emergence through agent-based methods allude to an organic process; they reflect a desire to bring the architectural process closer to one in which structure and form are interdependent. The rapid pace of development and adoption by the architecture community of various experimental software and workflow models – such as the traer-physics library for Processing or the Grasshopper plugins Kangaroo and Geometry Gym–are telltale signs that the computer is no longer seen by architects as merely a mechanism for representation divorced from physical conditions. Already, structural performance is part of the design process because of a traditionally close dialog between architects and structural engineers. Still, the challenge inherent to the creation of a truly generative computational tool for architectural design

---

Chris Bardt · Michal Dziedziniewicz · Joy Ko
Department of Architecture, Rhode Island School of Design, Providence, Rhode Island

remains, which is to establish materials and forces as agents of feedback in a dynamic way while modeling these processes accurately enough for the application at hand.

In this paper, we consider ribbed structures and explore strategies for rib growth in direct response to materials and forces. We have developed a tool and workflow that allows the structure to react, and to grow reactively. This digital "sandbox" integrates existing software – the 3D modeler Rhinoceros and the finite elements solver ABAQUS, software platforms that have widespread use in US architecture and engineering schools, respectively – putting an engineer and an architect in a position to start sharing platforms. Such a tool does not replace the engineer but has the potential to strengthen the architect, contributing to the architect's grasp of factors that influence structural performance at the early design stage. We demonstrate the methodological framework to set up an experiment using the problem setting of a gravity-loaded sheet of isotropic material and uniform thickness with a single point of support and propose a simple strategy for rib growth. This includes the calibration of parameters that can influence the quality of the experiment and can be used as a basis for a comparative study that would be difficult to do using existing tools. For the architect, the ability to access and gain awareness about performance drivers and to conduct meaningful experiments at an early stage opens up the opportunity for entirely different design strategies.

2 An Integrated Tool and Workflow

The tool described here is not a broad spectrum software, but rather the base for a family of specific applications. A representative application in this family is the problem of growing a pattern of ribs on a square plate supported at a number of points with a specified load towards some design objective of improved structural performance, such as maximum global stiffness of the plate. This tool supports the need for a workflow allowing users to experiment with growth rules based on design objectives and analytical feedback, and to see subsequent additions to the form. As such, this tool is not intended for a customized approach that takes a preexisting geometry (e.g. a pattern of ribs) as a starting point and morphs the geometry (e.g. such as thickening and thinning ribs) to improve on its performance.

The primary intended user-base for this tool is architects and architecture students, so the use of an existing, familiar, environment such as Rhinoceros was a priority. It is not realistic to expect an architect to fully comprehend and have the facility to be able to implement structural analyses classically conducted on this class of problems. However, a number of structural analysis platforms are now accessible to the practitioner who may have a good grasp of the fundamental principles underlying the modeled process but has limited to no understanding of numerical modeling. ABAQUS – a commercial finite elements analysis package that is popular amongst US engineering schools and select practices – is such a platform, but is not a magic "black box", and requires at minimum for the user to understand what is needed to generate the simulation: seed number and meshes, loads and boundary conditions, material properties. This amount of additional information is very reasonable to grasp for the architect, essential to avoid simulation errors, and can lead to more effective interdisciplinary collaborations.

The ultimate goal is to achieve a completely integrated tool in which ABAQUS is a plugin and Rhinoceros acts as the only operating environment. Currently our tool links Rhinoceros and ABAQUS – which have different base languages – through an independent "master" routine which communicates to each software through its command line batching protocol (Fig. 1). With the release of Python-based Rhinoceros 5.0 we hope to breach the barrier between languages.

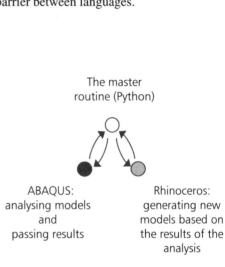

Fig. 1 An independent master routine (in Python) acts as the intermediary between VBscript-based Rhinoceros for geometry generation and Python based ABAQUS for the analysis, and calls out to either side in the generation process while also archiving and interpreting data. A set of templates have been developed that the master can rewrite into scripts to be used by either side of the process

While the role of Rhinoceros is primarily geometry-modeling, and that of ABAQUS is primarily analysis, there are a number of overlapping functionalities in the two software and subsequently a number of ways to distribute functionality responsibilities, with varying results. Where equivalent methods are present, ease of use and the establishment of a clear dividing line between geometry and analysis roles should govern. These were the guiding principles in the development of our integrated tool.

Rhinoceros handles all the geometric operations within the process while keeping the variants in the realm of so called Boundary Representation (Open Nurbs native geometry definitions) which are converted into ABAQUS-importable IGES files. For growing ribs of fixed cross section, a customized curve overlap, offset and extrusion routine were developed in place of the built-in solid body Boolean functions. Additionally, Rhinoceros responds to and records the list of growth nodes, which in turn determine the number of variants to be output and passed to the master routine alongside the IGES files for testing.

In ABAQUS, boundary conditions, meshing and material properties need to be specified for a simulation to be performed. Since the location and geometry of each support (in our case, a support "stem") is fixed in generating a given growth pattern, these supports are modeled within ABAQUS requiring no change to the ABAQUS script within a given run. Meshing is done using a tetrahedral meshing to handle forms that are not constrained to a predetermined grid. This is done via a specification of a seeding of the boundary, which in turn is a function of the seed number (an average node-vertex to node-vertex distance). By fixing the seed number we fix the nodes on the surface and hence, obtain a consistent set of vertices throughout a run. For a design objective of least square deflection of the plate, these plate vertices are further identified by looking at the coordinates of each vertex of the instance created and matching the vertices in the sheet with corresponding field outputs. By using ABAQUS, there is the advantage of an extensive material library which is constantly expanding through such additions like the Granta material selection plugin.

3 Case Study: A Design Experiment to Model Rib Growth

3.1 Historical Significance of Rib Forms

The relation of ribs to surface form has intrigued architects and engineers for hundreds of years. The early medieval masons developed ribs based on drawn arcs of circles, and the vaults spanning these ribs were distorted and uneven irregular surfaces "stretched" to fit between the splayed rib structures [9]. The ribs served as the structure, to a large extent, carrying the vaults. The 16th century development of sophisticated drawing systems allowed masons to control the stonecutting of vaults to such an extent that the vaults became shell structures and the ribs a reinforcing lattice work – in effect reversing the structural role of vaults and ribs from the early Gothic period [6].

Historically, there has been a fundamental ambiguity between rib and surface. Which comes first – the form of the rib or the the surface which is being reinforced by the rib? In the 20th century, rib and surface research was conducted by a new breed of designer, the architect-engineer such as Nervi, Maillart, Dieste, Candela. Their research took the form of experimental long span structures such as thin shells, ribbed shells, and lamella structures often using the new technology of reinforced concrete. These experiments were limited to statically determinate structures, geometries optimized for given parameters. Any kind of experimentation that went beyond these structures proved difficult, an example being Frank Lloyd Wrights tapered, dendriform, mushroom columns which were subject to a combination of tensile, coplanar and non axial loads. The local building committee refused to approve the column for construction until its performance had been empirically demonstrated [8].

In recent decades, the problem of generating optimal patterns of reinforcement in plate and shell structures – topology optimization – has been studied intensively by the mathematical and engineering communities. Some notable precedents have informed our work, including Bendsoe and Kikuchi's work [2] which spawned studies of a large range of loading conditions, optimality and efficiency criteria and hybrid materials (many detailed in [3]) that utilize the homogenization method–a powerful tool in variational calculus–which results in solutions that are density distributions of material. For practical concerns of buildability and formwork, we looked to techniques that produced distinct pattern of reinforcements. In [5], Ding and Yamazaki proposed a technique based on the adaptive growth rule of branching systems using a design criterion of maximum global stiffness, which has the benefit of producing distinct patterns. While it seems quite general to a number of support and loading conditions, only typical support and loading conditions on a square plate were implemented.

3.2 Creating the Methodological Framework for an Experiment

The integrated tool and workflow is designed as a sandbox in which experiments to test intuition-led strategies can be conducted with relative ease. In devising this initial experiment, a number of assumptions were made. The surface that we consider in this experiment is a square plate with uniform thickness and a single circular column of support. The ribs are uniform in cross-section and of fixed length. The surface and ribs are made of the same isotropic elastic material and form a monolithic ribbed surface. The boundary conditions applied are total encastre for the support allowing displacement of the ribbed surface. Since the surface here is a plate, we can consider a design objective of least square deflection of the surface. The set of mesh nodes on the plate, $\mathcal{N}$, stays constant through a given run so we can define deflection in the least-squares sense by $\sqrt{\Sigma_{i \in \mathcal{N}} \Delta z_i^2}$, where Δz_i is the vertical deflection of the ith node relative to the height of the plate above the support. The growth strategy used here is based on a "greedy algorithm" approach where at each step in the iteration, the choice that minimizes the squared deflection amongst all local possibilities is taken. This growth strategy comprises the following rules:

- Growth initializes at the position of the support, which we call the origin.
- An active growth point is either the end of a connected set of ribs emanating from the origin or the origin itself.
- At each active growth point, there are k equivariant possible directions of growth. At all active growth points other than the origin, the length of the rib grown at each step is a fixed length l. The length of a rib emanating from the origin is length $r+l$, where r is the radius of the circular support.
- A finite elements analysis for the whole structure is carried out at each growth point for each direction of growth, and rib growth occurs at the point and direction that minimizes the least squared deflection amongst all local possibilities.
- When the rib hits the boundary of the plate, the rib is cut short and the point of intersection with the boundary becomes an active growth point.

This initial experiment is intended to be interactive with the user so no a priori stopping condition is given other than the number of steps that the user specifies, or when the ribs at all active points grows back onto itself. Fig. 2. illustrates the rules governing this growth strategy.

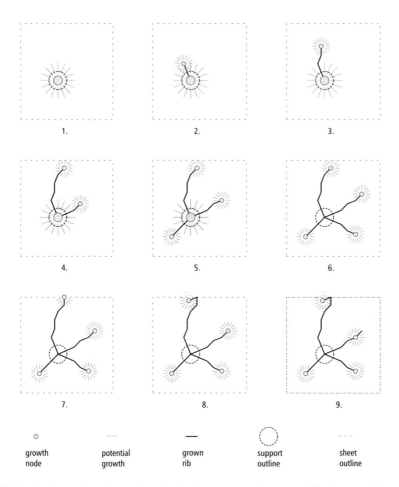

Fig. 2 Configurations along a hypothetical pattern growth governed by the growth strategy

A number of parameters can influence the effectiveness and accuracy of the calculation of rib growth according to this growth strategy. These in general depend on the study at hand; in our case, we wished to conduct a comparative study in which the position of the support moves along the diagonal of the plate (Fig 3). Diagnostics were run to determine a choice of plate dimensions, stem radius, material properties and seed number so that the maximum deflection of the plate supported at each point being considered was sufficiently small and so that the results could be meaningfully compared. The dimensions of the rib unit can greatly influence the effectiveness of the growth; too high a volume increment and the rib can easily increase the deflection in the plate; too low a volume increment and the iterations in growth typically reduce and meaningful growth patterns may not be obtained. Since our interest is the pattern of growth, we focused on the influence parameter of the depth of the

Tools and Design Strategies to Study Rib Growth

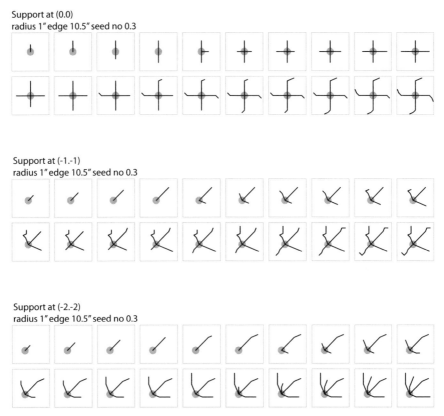

Fig. 3 A comparative study showing 20 iterations of rib growth corresponding to the proposed growth strategy starting with a support at the center and moving out towards the diagonal. Parameters for these runs: plate parameters given by edge length 10.5 inches (26.67 cm) and depth of 1/8th inch (0.3175 cm); stem radius of 1 inch; material properties given by Young's modulus of 69 GPa and Poisson's ratio of 0.3; seed number of 0.3; rib parameters with length of 1 inch (2.54 cm), thickness of 1/8th inch (0.3175 cm) and depth of 1 inch (2.54 cm); 16 directions of growth at each node.

rib, fixing the length and the thickness. Fig. 4 shows three choices of rib length of a choice of support position corresponding to the position of the support offset at from the center of the plate. The monotonically decreasing deflection curve corresponds to continued growth of ribs leading to a meaningful growth pattern, whereas a flattening out corresponds to no further growth.

From this initial experiment, a natural evolution of the proposed growth strategy is one based on a variable volume increment which might reveal structure at a finer scale and would be a natural quantity on which a stopping condition could be based. Additionally, initiating growth at multiple points with different rates of growth could be a more effective strategy to cover more ground with less material intensity of ribs.

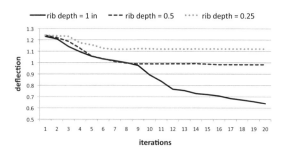

Fig. 4 Deflection, in the least squared sense, corresponding to each iteration of a run for three choices of rib depth for the case where the center of support is offset from the center of sheet by $(-1,-1)$. Remaining run parameters are the same as those used for the comparative study Fig 3.

4 Conclusion

The integrated tool and workflow provides a digital sandbox in which experiments to test intuition-led strategies on a class of problems, including ones on rib growth, can be conducted. By interacting with the structure, the architect gains valuable awareness to structural factors which can inform design decisions.

Acknowledgements. We would like to thank Viswanath Chinthapenta for launching us on our journey with ABAQUS, to Shane Richards for lending us his time and expertise in our parallel journey in fabrication, and to the Brown University Engineering School for granting us access to their Computational Mechanics Research Facility.

References

1. Banham, R.: Theory and Design in the First Machine Age. MIT Press, Cambridge (1980)
2. Bendsoe, M.P., Kikuchi, N.: Generating optimal topologies in structural design using a homogenization method. Comput. Methods Appl. Mech. Eng. 71, 197–224 (1988)
3. Bendsoe, M.P., Sigmund, O.: Topology Optimization, 2nd edn. Springer, Heidelberg (2004)
4. Le C.: Towards a New Architecture. John Roder, London (1931)
5. Ding, X., Yamazaki, K.: Adaptive growth technique of stiffener layout pattern for plate and shell structures to achieve minimum compliance. Engineering Optimization 37(3), 250–276 (2005)
6. Evans, R.: The Projective Cast, Architecture and Its Three Geometries. MIT Press, Cambridge (1980)
7. Hearn, M.F. (ed.): The Architectural Theory of Viollet-le-Duc, Readings and Commentary. MIT Press, Cambridge (1990)
8. Lipman, J.: Frank Loyd Wright and the Johnson Wax Buildings, Rizzoli (1986)
9. Willis, R.: On the Construction of the Vaults of the Middle Ages. Royal Institute of British Architects, London (1842)

Free Shape Optimal Design of Structures

Kai-Uwe Bletzinger

Abstract. Actual trends in numerical shape optimal design of structures deal with handling of very large dimensions of design space. The goal is to allowing as much design freedom as possible while considerably reducing the modeling effort. As a consequence, several technical problems have to be solved to get procedures which are robust, easy to use and which can handle many design parameters efficiently. The paper briefly discusses several of the most important aspects in this context and presents many illustrative examples which show typical applications for the design of light weight shell and membrane structures.

1 Introduction

Shape optimal design is a classical field of structural optimization. Applied to the design of free form shells and membranes or, more generally, light weight structures, it is of big importance in architecture, civil engineering or various applications of industrial metallic or composite shells as e.g. in automotive or aerospace industries [1, 2, 3]. In the "old" days of the pre-computer age optimal shapes had been found by experiments such as inverted hanging models or soap film experiments. Still, those shapes are of great importance for practical design as they define structures of minimal amount of bending which, in turn, are as stiff as possible. As a consequence, "stiffness" is one of the most important design criteria one can think of. The methods discussed in the sequel refer to this design criterion in various ways.

A standard approach of optimal shape design is to discretize the structure and to use geometrical discretization parameters as design variables, e.g. nodal coordinates. As optimization is a mathematical inverse problem it exhibits typical pathological properties which in particular become obvious or even dominant if the number of design parameters becomes large. In particular, one has to deal with questions like

---

Kai-Uwe Bletzinger
Lehrstuhl für Statik, Technische Universität München, Germany

irrelevant degrees of freedom tangential to the surface, highly non-convex design spaces, and mesh dependency, just to mention the most important. The state-of-the-art answer to those problems is to use CAGD methods for the discretization of geometry: The success of that approach, however, is a consequence of the reduced number of design parameters rather than a consequent elimination of the source of deficiencies. In other words, if the number of CAGD parameters used for structural optimization is increased, the pathological properties become obvious, again.

If geometrical parameters of a fine discretization are used, as e.g. the coordinates of a finite element mesh, strategies have to be developed to stabilize the original deficiencies of the inverse problem.

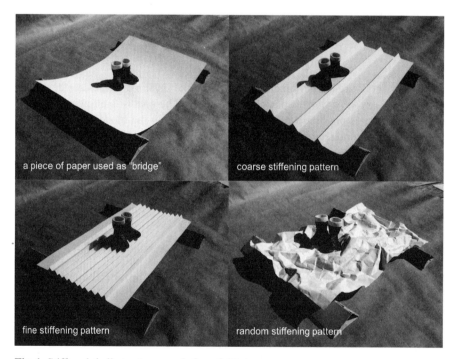

Fig. 1 Stiffened shell structures made from folded paper

2 Design Noise and the Infinity of Design Space

The principal challenge of form finding can briefly be explained by an illustrative example. The task is to design the stiffest structure made from a piece of paper which is able to act as a bridge carrying load. The solution is well known. As the piece of paper is unable to act in bending stiffeners have to be introduced by folding the paper. However, there exists an infinite number of solutions which all of them do the job creating stiff solutions of at least similar quality which is by far better than the quality of the initially flat piece of paper. Surprisingly enough, even an arbitrary pattern of random folds appears to be a possible solution, Fig.1. The figure of the randomly crinkled paper is an ideal paradigm for the infinity of the design space or, more ostensive, the "design noise". As for the actual example the crinkled paper can be understood as the weighted combination of all possible stiffening patterns one can easily think of a procedure to derive any of the individual, basic solutions of distinct stiffening patterns by applying suitable "filters" to the design noise. It is clear that the kind of "filter" as well as the "filter process" can be freely chosen as an additional and most important design decision. It is possible to define a procedure as implied by the actual example and to, first, generate a "highly frequent" design noise and to apply geometrical filters in a second step. It is, however, also possible to apply "indirect" filters by preselecting and favoring certain classes of solutions in advance. There is no doubt that the mentioned second way is the more ingenious one as a large set of other, perhaps even better solutions, might be undetected. It remains to the insight of the applying per-son about how to define a procedure of pre-selection, regularization or "pre-filtering", just to refer to the introduced picture. Most often, however, there remain some secrets or at least some vagueness. From this point of view form finding truly is an art.

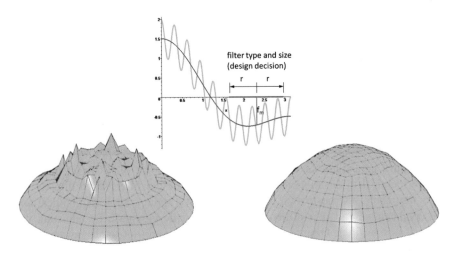

Fig. 2 Direct numerical stiffness optimization and filtering of a plate subjected to self-weight

3 Direct Filtering of Numerical Models

Numerical form finding techniques allow the direct approach of filtering the design noise. As an example, consider a circular plate as shown in Fig.2 which is discretized by finite shell elements. The nodal positions of the finite element nodes shall be found such that the stiffness of the structure is maximized. Without any additional treatment numerical optimization procedures suggest a highly crinkled shape. Obviously, the simulation resolves the physics of the random stiffening pattern similar to the paper experiment. The smallest crinkles are defined by the possible resolution of the finite element mesh. Finer meshes allow even higher frequencies. Additionally, high frequent crinkles result in extreme element distortions which come together with additional artificial, non-physically stiff element behavior which is known as "locking". That means that filtering play a double role (i) to prevent non-physical artifacts by controlling mesh distortion and (ii) to help to identify the preferred optimal shape within the infinity of the design space. For the plate in Fig.2 a coarse low pass filter has been applied. It is a simple hat function of rotational symmetry with a base length of the size of the plate diameter. Consequently, the dome is identified as optimal structure which is well known from the inverted hanging model experiment. It appears that the type of filter (e.g. hat, cubic spline or Gauss distribution) is of minor importance in contrast to the size of the filter basis which directly controls the minimum size of stiffening "crinkles". The filter size is a very effective as well as efficient control for exploring the design space as it is shown in the following examples [1, 2, 3, 4].

4 CAGD Based Parameterization Techniques and Structural Shape Optimization

The industrial state of the art in structural optimization is characterized by the combined application of CAGD methods *(Computer Aided Geometric Design)*, finite element analysis, and non-linear programming. The idea is to define the degrees of freedom for shape optimization and form finding by some few but characteristic control parameters of the CAGD model. The choice of a CAGD model is indeed identical to an implicit pre-selection of a design filter which directly affects the result. As most often a CAGD model is quite complex, modifications are cumbersome and it is difficult to explore the design space by adjusting the implicit design filter. Often architects and engineers are not totally aware about that and miss alternatives. Still, however, the remaining design space might be large enough and the limitations might be accepted. The most actual trend is defined by the Isogeometric Analysis, where NURBS shape functions are used for both, the design modeling as well as the structural analysis [5, 6, 7, 8, 9].

As a consequence, design models must be analysis suitable which create new challenges for the CAGD community regarding geometrical compatibility and treating trimmed surfaces. T-splines have been suggested as remedy [10, 11].

5 Minimal Surfaces

The form finding of tensile structures is defined by the equilibrium of external and internal pre-stress forces. The choice of pre-stresses of surface and edge cables is the "filter" applied to screen the design space. As the shape is uniquely defined by the equilibrium of forces and stresses there is no material related term in the equations. Consequently, nodes of the discretization mesh can float freely on the surface because surface strains are not inducing elastic stresses relevant for the form finding process. Additional regularization of the method is necessary for procedural reasons not as means to explore the design space. In contrast to the most of the available methods the Updated Reference Strategy (URS) is consistently derived from continuum mechanics [12, 13]. Therefore, it appears to be very robust and can easily be applied for all kind of applications, for membranes as well as cables and their combinations. It can be interpreted as a generalization of the well-known force density method [14].

6 Illustrative Examples

6.1 Pre-stressed Surfaces

These examples, Fig.3 and Fig.4, present the direct application of URS for the design of pre-stressed surfaces due to isotropic (minimal surfaces) and anisotropic surface stresses. Note, that even ideal minimal surfaces can easily be determined which is a challenge for many available structural form finding methods. The implemented procedure is able to treat form finding under additional effects as there are additional surface loads (e.g. pressure), interior edge cables (needs additional formulation of constraints on cable length) and consideration of stiffening members in bending and compression (kind of tensegrity structures). For further information refer to *www.membranes24.com*.

6.2 Norwegian Pavilion at EXPO 2010, Shanghai

This example shows the application of the URS technique in architecture and civil engineering for the form finding of the roof for the Norwegian pavilion at the EXPO 2010 in Shanghai, Fig.5, [15].

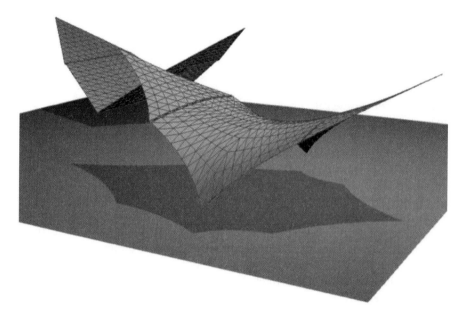

Fig. 3 "Bat Wing", Form finding of hybrid structure: Isotropic surface stress, edge cables, spokes in compression and bending

Fig. 4 Form finding of minimal surfaces and ideal spherical soap bubbles

6.3 Bead Design of Plates and Shells for Single Loads

A bend cantilever made of a thin (metal) sheet is loaded as shown, Fig.6. A filter radius as large as the width of support is used. The model consists of appr. 5.000 shape variables. The optimal shape (most right) is reached after 19 iteration steps.

Free Shape Optimal Design of Structures

Fig. 5 The roof of the Norwegian pavilion at 2010 EXPO, Shanghai: Application of URS

Another example demonstrates the mesh independence of the method, Fig.7. A quadratic plate is loaded in the center and supported at the corners. The question is to find the optimal topology of stiffening beads. A filter radius is chosen as large as half of the width of support. Additionally, a constraint on the maximum bead depth is given. As shown, the optimal solution is characterized by the filter but it is mesh independent. The choices of filter type and size are additional degrees of design freedom which may be used to explore the design space. Note the smooth final surface although local radial filters are applied.

Fig.8 shows the result of a joint project together with Adam Opel GmbH. The optimal distribution of beads has been determined to maximize the five lowest eigenfrequencies of a thin metal sheet. The number of iterations appears always to be not more than 40 for every problem size.

Fig. 6 Shape optimization of a cantilever shell

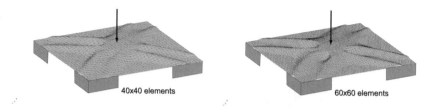

Fig. 7 Optimal bead design of initially plane sheet

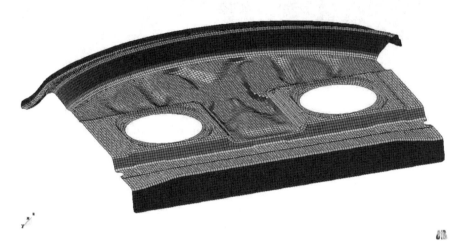

Fig. 8 Bead optimization of a thin metal sheet for the automotive industry

6.4 Shape Optimization of a Wind Turbine Blade

The shape of a wind turbine blade is optimized for two cases, Fig.9 to Fig.11: To maximize stiffness for given mass and to minimize mass for defined stiffness. The pressure distribution has been determined from a CFD simulation is applied to a linear elastic structural model for shape optimization as a preliminary design study. The next steps will consider a complete non-linear structural model in a fully coupled FSI-environment for shape optimization. More than 9.000 shape variables have been used. Again, note, the smooth shape although small design filters have been used to prevent numerical noise. The initial shape has been generated from a Rhino 3D ©model which also can be used for a isogeometric analysis for the fully coupled, transient analysis of the blade in a numerical wind tunnel [7, 8, 9]. The latter study has been done in a joint work together with Yuri Bazilevs at University of California at San Diego [16].

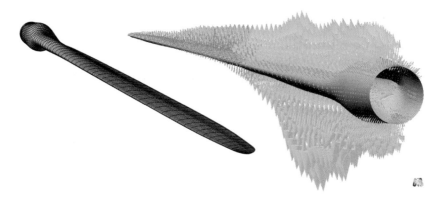

Fig. 9 Rhino 3D model of wind turbine blade (left); wind pressure distribution (right)

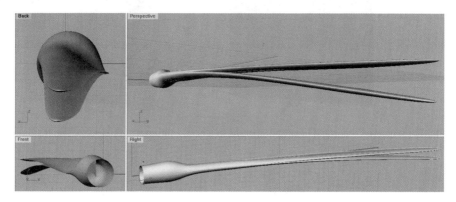

Fig. 10 Screen dump of the Rhino-Plug-In developed at Lehrstuhl für Statik used as pre- and post-processor for isogeometric design and analysis of non-linear shell structures

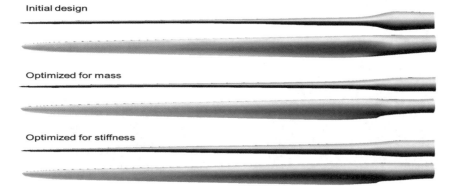

Fig. 11 Optimized shapes from a free mesh, filter based optimization procedure

6.5 Staggered Optimization of a Fiber Reinforced Composite Shell

The shape of the bend cantilever is again determined, now assuming a composite shell with two layers of fiber reinforcement, Fig.12 and Fig.13. The filter technique has been applied to regularize the fiber optimization as well. The objective

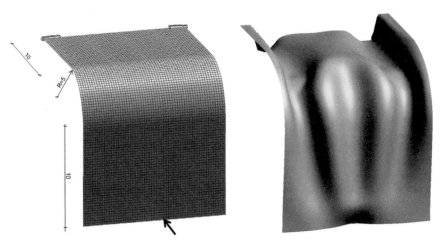

Fig. 12 Staggered shape and fiber optimization of a bend cantilever; Initial shape and loading (left), optimal shape (right)

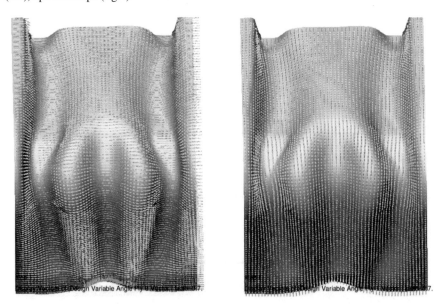

Fig. 13 Staggered shape and fiber optimization of a bend cantilever; Optimal fiber orientation, bottom layer (left), top layer (right)

Free Shape Optimal Design of Structures 35

is maximum stiffness; altogether there are about 80.000 shape and fiber angle variables. Problems like this need most sophisticated simulation techniques, including highly efficient semi-analytical [17] and adjoint sensitivity analysis, robust, gradient based optimization techniques (variants of conjugate gradients), reliable finite element models as well as efficient, object-oriented and parallel implementation in the institutes code CARAT++.

6.6 Design Chain

This example demonstrates the design chain of a shell structure from a Rhino design model, via a finite element analysis and optimization model to the final optimized shape for stiffness under self-weight applying the mesh based filter-technique which allows for the largest possible design space. Fig.14 and Fig.15

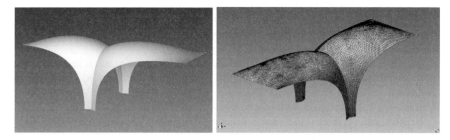

Fig. 14 Rhino design model (left), FE model for analysis and optimization (right)

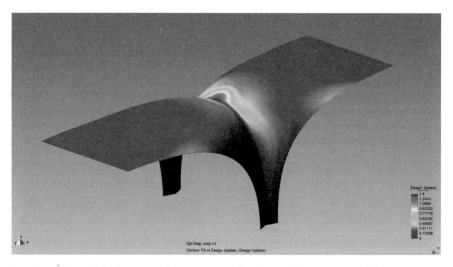

Fig. 15 Design chain: Design update contour indicating modification potentials at the gorge. The number of optimization variables is correlated with the number of finite element nodes allowing for an extremely large design space, much larger than the NURBS design model.

Acknowledgements. The funding of the Deutsche Forschungsgemeinschaft, International Graduate School for Science and Engineering at TUM as well as the European Union are gratefully acknowledged. Also, all the members of the Lehrstuhl für Statik should be mentioned who have contributed to this work over the years. Among them shall be mentioned as representatives for all: Michael Breitenberger, Natalia Camprubi, Ferna Daoud, Matthias Firl, Michael Fischer, Majid Hojjat, Helmut Masching, Josef Kiendl, Johannes Linhard, Robert Schmidt, Electra Stavropoulou and Roland Wüchner.

References

1. Bletzinger, K.U., Firl, M., Linhard, J., Wuchner, R.: Optimal shapes of mechanically motivated surfaces. Computer Methods in Applied Mechanics and Engineering 199, 324–333 (2010)
2. Bletzinger, K.U., Wuchner, R., Daoud, F., Camprubi, N.: Computational methods for form finding and optimization of shells and membranes. Comp. Methods Appl. Mech. Engrg. 194, 3438–3452 (2005)
3. Linhard, J., Wuchner, R., Bletzinger, K.U.: "Upgrading" membranes to shells. The CEG rotation free shell element and its application in structural analysis. Finite Elements in Analysis and Design 44, 63–74 (2007)
4. Wuchner, R., Firl, M., Linhard, J., Bletzinger, K.U.: Plateau regularization method for structural shape optimization and geometric mesh control. PAMM 8, 10359–10360 (2008)
5. Hughes, T.J.R., Cottrell, J.A., Bazilevs, Y.: Isogeometric analysis: CAD, finite elements, NURBS, exact geometry, and mesh refinement. Computer Methods in Applied Mechanics and Engineering 194, 4135–4195 (2005)
6. Cottrell, J.A., Hughes, T.J.R., Bazilevs, Y.: Isogeometric Analysis: Toward Integration of CAD and FEA. John Wiley & Sons, Chichester (2009); ISBN 978-0-470-74873-2
7. Kiendl, J., Bletzinger, K.U., Linhard, J., Wuchner, R.: Isogeometric shell analysis with Kirchhoff. Love elements. Computer Methods in Applied Mechanics and Engineering 198, 3902–3914 (2009)
8. Kiendl, J., Bazilevs, Y., Hsu, M.C., Wuchner, R., Bletzinger, K.U.: The Bending Strip Method for Isogeometric Analysis of Kirchhoff-Love Shell Structures Comprised of Multiple Patches. Computer Methods in Applied Mechanics and Engineering 199, 2403–2416 (2010)
9. Schmidt, R., Kiendl, J., Bletzinger, K.U., Wuchner, R.: Realization of an integrated structural design process: analysis-suitable geometric modeling and isogeometric analysis. Computing and Visualization in Science 13, 315–330 (2010)
10. Sederberg, T.W., Zheng, J., Bakenov, J.A., Nasri, A.: T-Splines and T-NURCCS. ACM Transactions on Graphics 22, 477–484 (2003)
11. Bazilevs, Y., Calo, V.M., Cottrell, J.A., Evans, J.A., Hughes, T.J.R., Lipton, S., Scott, M.A., Sederberg, T.W.: Isogeometric analysis using T-splines. Computer Methods in Applied Mechanics and Engineering 199, 229–263 (2010)
12. Bletzinger, K.U., Ramm, E.: A general finite element approach to the form finding of tensile structures by the updated reference strategy. Int. Journal of space structures 14, 131–146 (1999)
13. Wuchner, R., Bletzinger, K.U.: Stress-adapted numerical form finding of pre-stressed surfaces by the updated reference strategy. Int. J. Numer. Methods Engrg. 64, 143–166 (2005)

14. Linkwitz, K.: Formfinding by the Direct Approach and pertinent strategies for the conceptual design of pre-stressed and hanging structures. Int. Journal of Space Structures 14, 73–88 (1999)
15. Lienhard, J., Abrahamsen, R., Schoene, S., Soto, M., Knippers, J.: The Norwegian Pavilion at the Expo Shanghai 2010. In: IASS Conference, Shanghai (2010)
16. Bazilevs, Y., Hsu, M.C., Kiendl, J., Wüchner, R., Bletzinger, K.U.: 3D Simulation of Wind Turbine Rotors at Full Scale. Part II: Fluid-Structure Interaction. International Journal of Numerical Methods in Fluids 65, 236–253 (2011)
17. Bletzinger, K.U., Firl, M., Daoud, F.: Approximation of derivatives in semi-analytical structural optimization. Computers and Structures 86, 1404–1416 (2008)

NetworkedDesign, Next Generation Infrastructure for Design Modelling

Jeroen Coenders

Abstract. This paper will discuss several key features which have been introduced as part of the new computational infrastructure for design modelling, called 'NetworkedDesign'. The paper will describe and illustrate these features and will present their underlying concepts by making use of a novel computational prototype system (under the same title) which the author has developed from the ground up.

NetworkedDesign aims to support the adoption of more and more advanced computational technology in (structural) design and engineering by attempting to introduce new concepts as well as by combining a number of existing modelling paradigms in a flexible and user adaptable system. This will allow for new opportunities to model and use rich sets of information, knowledge and experience which can be reused to enhance the designs of the future.

1 Introduction

During the life cycle of a building or structure, especially during the design and engineering phases, a lot of information and data is being produced of which a large proportion is lost due to the lack of computational infrastructure that supports these processes close enough to collect and reuse this information in a meaningfull and manageable way. A large potential exists for this information to be mined at a later stage to learn and produce better designs in the future. Furthermore, a range of advanced technologies exists in the form of analysis, simulation, modelling, optimisation, digital fabrication, etc. which are currently hardly used during the production of a building (design) and which could further improve designs.

Jeroen Coenders
Arup, Amsterdam, The Netherlands
BEMNext Laboratory, Faculty of Civil Engineering and Geosciences,
Delft University of Technology, Delft, The Netherlands

To reach a higher level of adoption of computation in the design and engineering profession and its processes, the designer and engineer should be empowered by the tools. They cannot be persuaded by force to adaopt these tools (which has happened with many of the previous approaches), but they have to be seduced to change. This research assumes that this will occur when ease of use, powerful features and a close relationship to the key values in design will seduce the user to use these methods.

This paper will discuss a conceptual computational infrastructure which has been designed based on user studies to provide a base for a new design process where the designer is able to use a combination of modelling paradigms, logic and flexibility to aid and support his or her design process so that rich information can be collected for later use. This paper will discuss several key features which have been introduced as part of this infrastructure called 'NetworkedDesign'. The paper will describe and illustrate these features, and will present their underlying concepts by making use of a novel computational prototype system (under the same title) which the author has developed from the ground up.

2 NetworkedDesign

NetworkedDesign is an infrastructure to build the tools, frameworks, applications, systems and standards of the future on. NetworkedDesign aims to support the adoption of more and more advanced computational technology in (structural) design and engineering by attempting to introduce new concepts which have a close relationship to the act, the process, the characteristics of and the core values in design as well as by combining a number of existing modelling paradigms in a flexible system under the assumption that if the user has a rich toolbox available for modelling, he or she will be able to use more computational technology in the design process. This will allow for new opportunities to model and use rich sets of information, knowledge and experience which can be reused to enhance the designs of the future. Below a number of the new concepts will be discussed. An extensive description of all concepts part of NetworkedDesign can be found in [2].

2.1 Novel Concepts

NetworkedDesign combines a range of concepts from other modelling systems, such as most concepts from parametric and associative design, rule-based modelling, object-orientation, etc. NetworkedDesign allows the user to select the preferred solver of choice (or a combination of solvers) to solve the logic which has been defined to describe reality in the model.

An example has been presented in Figure 1 where a parametric modelling is combined with a rule-processing approach to build a powerful conversion tool to quickly convert any geometrical model in a structural model by selection and filtering of the relevant objects.

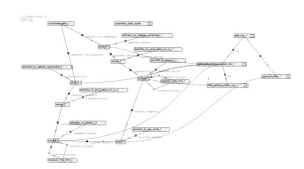

Fig. 1 Example of the application of rule-processing logic. On the left-hand side of the logic the geometric definition of a simple twisted tower has been built. On the right-hand side a separate rule-based logic can be seen which selects and filters certain elements from the geometric model to be used for instance for a structural analysis. The blue lines indicate ad-hoc created relationships by the rule-processing engine between the relevant rule and the objects it references

Below a selection of the novel concepts introduced in NetworkedDesign will be illustrates.

2.1.1 Multiple Method Types

NetworkedDesign contains objects in the form of containers of properties and behaviour (open objects). These are added by making use of methods of which multiple types and instances can be combined to create a custom definition of an object in reality.

An example can be seen in Figure 2 where several method types have been combined in a single model, which will be a usual case for NetworkedDesign models. The model defines a simple line by two points. The points have been combined into a single definition in Cartesian space (by providing a coordinate system and X, Y and Z coordinates) by making use of replication (which is a concept for parametric and associative design modelling in GenerativeComponents by Robert Aish [1]). These points have been transformed from their Cartesian defintion into a spherical coordinate system. Transformation is a concept which will be further discussed in Section 2.1.2. The points have been assigned to a line as a begin and end point of the line. Finally, a calculation of the length of the line is performed as an additional property on the line object.

The different method types combined in this example model are: definition, transformation and calculation. Definition allows the user to define an object by providing property values to an object, such as the X, Y and Z coordinates in the example. Calculation allows for the addition of new properties and providing values to these properties based on a predefined calculation procedure (such as the calculation of

Fig. 2 Example of some of the different method types in NetworkedDesign

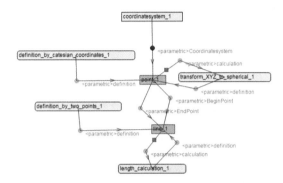

Fig. 3 Example of transformation

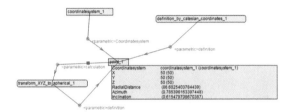

the length of a line). Each time the object is updated with new values, the calculation will re-execute to determine these values again for the new object properties.

2.1.2 Transformation

Transformation allows the user to bi-directionally transform the definition of an object into another definition with the option of additional input of new information (not used in the example). This involves querying for new information, taking existing properties from an object, testing the properties and transforming by making use of a predefined procedure.

In Figure 3 the result of a transformation can be seen. Like the previous example the point has been transformation from a Cartesian definition to a spherical coordination system. In the property list of the object first the four properties can be seen which have been added by the Cartesian definition: the base coordinate system, the X, Y and Z coordinate. The next three properties have been added by the transformation to a spherical coordinate system: the distance to the origin (radial distance), the azimuth and inclination angle.

2.1.3 Bi-directionalty and Solution Finding

Because design problems often contain bi-directional relationships, Networked-Design has been designed to support these logical relationships. In most other

Fig. 4 Example of a bi-directional problem. The original parametric definition adds two numbers (doubles) and adds a third number to obtain the final outcome. By declaring the first input number a variable and by declaring a goal value on the outcome, the bi-directional problem is defined: how to modify the first input so that the goal value becomes the outcome?

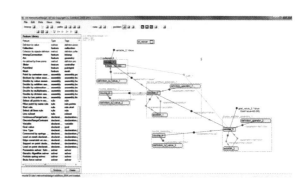

modelling systems that allow modelling of logic (in contradiction to placement and manipulation of geometrical objects which most CAD systems do) support for bi-directionality is non-existent or very restrictive. For example, parametric and associative modelling systems only allow for input-to-output relationships to build a logical graph. However, this graph cannot reconnect on itself. Bi-directionality can in that case be simulated by repetitive update loops by an external controller, but this provides rather limited possibilities. In other systems, such as constraint modelling systems, the logical relationships do not have be ordered in a graph, but are directly solved by a constraint solver which solves a system of equations. Bi-directional relationships are allowed in these systems. The problem that occurs is that higher order equations are difficult to solve and therefore these systems are usually limited to simple geometrical objects, such as points, lines and circles, or have limited abilities to transform objects.

Figure 4 and Figure 5 show one way how NetworkedDesign can deal with bi-directional problems (simulation by repetitive loops) and Figure 6 illustrates another way (reversion of logic). NetworkedDesign can also employ solvers, such the constraint modelling solvers, which directly solve the logic.

Note that another mechanism of NetworkedDesign is in play in these figures: solution finding. The user is able to express a problem to which he or she does not know the solution and the system can search through a database of predefined solutions and solution methods to suggest a solution to the problem to be applied.

2.1.4 Multi-directionality

Because NetworkedDesign is not limited to single definitions or one single solution mechanism, it can occur that a problem could be solvable with a variety of methods. For this case NetworkedDesign has introduced the concept of multi-directionality, so that multiple paths through the logic can be used to resolve the logic. Figure 7 illustrates multi-directionalty and its differences to single and bi-directionality.

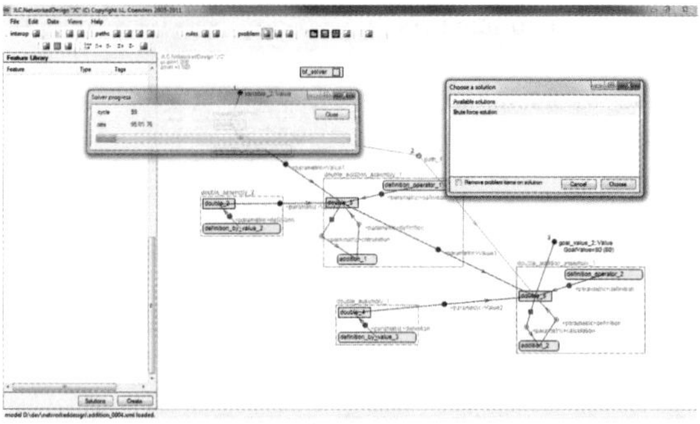

Fig. 5 Example of a solution of the bi-directional problem presented in Figure 4. On the right-hand side the system has presented the possible solutions it knows to the presented problem: in this case application of a brute-force solver (trying all permutations). The screenshot has been taken after selection of this solution and on the left-side the running solver can be seen in action.

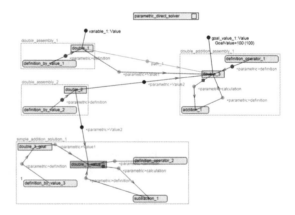

Fig. 6 Example of a solution of a similar but simpler bi-directional problem where two numbers are added to reach the outcome. The logic of this problem can be reversed by making use of subtraction in additional piece of logic that defines this bi-directional relationship. It must be mentioned that a small limitation in the current implementation prevents direct integration in the original logic, leading to copied objects for the original numbers. This will be resolved in a future version of the prototype system.

2.2 Implementation

On top of a prototype implementation of the 'NetworkedDesign' infrastructure a prototype design system has been implemented under the same title to demonstrate and test some of the capabilities of the infrastructure. This prototype has not been implemented for use by end-users as no usability, interaction or user interface design has taken place, but contains a large portion of the concepts introduced. The current implementation has been built on top of the .NET framework provided by Microsoft[1] and has been implemented in C# as programming language. The infrastructure is not

---

[1] http://www.microsoft.com/net

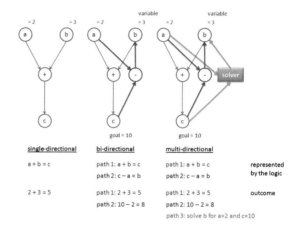

Fig. 7 Examples of single-directionality, bi-directionality and multi-directionality in a schematic graph.

limited to this framework or language, but these have been chosen as many of the existing design modelling systems allow extensions in the form of plug-ins, add-ins or COM/.NET interface interoperability. The current system implementation can be used as a stand-alone system, a plug-in which can run in Rhino3D by McNeel & Associates[2] and is able to interoperate with a multitude of systems, such as Bentley Systems' Microstation[3], Tekla Structures[4], Oasys GSA[5] and Microsoft Office[6]. All figures in this paper, except Figure 7, are screenshots or are based on screenshots of the prototype modelling system which has been built on top of the 'NetworkedDesign' infrastructure.

3 Discussion

The infrastructure described in this paper should be seen as a first step towards a computational infrastructure to support the future of design and engineering. It needs to be noted that although the first steps have been made, the introduction of this new infrastructure needs to be further researched and potentially expanded with other concepts found in modelling systems. Furthermore, the infrastructure will have to reach complete implementation and new design systems will need to be build on top of the infrastructure before the value of the infrastructure can be fully studied.

Over the last 20 to 30 years Building Information Modelling or BIM [3] has been developing in the industry with similar goals to become an infrastructure for the building life cycle. So far the approach most BIM software applications have

---

[2] http://www.rhino3d.com
[3] http://www.bentley.com/en-US/Products/MicroStation+Product+Line/
[4] http://www.tekla.com
[5] http://www.oasys-software.com/products/structural/gsa
[6] http://office.microsoft.com

been taking has not succeeded in modelling design and engineering close enough to achieve this goal. NetworkedDesign aims to provide a more enhanced approach to model on the level of logic rather than objects. Furthermore, the NetworkedDesign infrastructure provides, by design, a route to attack the interoperability (the ability of systems to exchange information with each other) issue other approaches suffer from. NetworkedDesign aims to attack this issue by including open objects and a number of mechanisms for the user to define his or her own interoperability templates, or to use a pre-defined standard. Open objects have been discussed in Section 2.1.1. Since the user can build any interoperability template himself or herself, any template for interoperability, including pre-agreed standards, can be used. The different method types and the properties they add to the system are useful components to build pieces of logic which could easily exchange with other systems, because the resulting objects will share a common logical definition. This makes NetworkedDesign very flexible and applicable to any situation: just like an infrastructure should be. Transformation and mechanisms like rule-processing provides enough flexibility to deal with many logical structures and the ability to transform one logic to the logic of another system. Agreement between these systems is not required, as the user is able to define himself or herself how the logic is transformed. Concepts of bi-directionality and multi-directionality provide the ability to build highly reusable templates for different systems and models.

4 Conclusions

In this paper the new 'NetworkedDesign' infrastructure for design modelling has been discussed which merges and extends a number of existing approaches, such as parametric and associative design modelling and rule-based modelling, and introduces several new concepts, such as multi-directionality. This infrastructure has implemented in a prototype system under the same title which demonstrates the concepts collaborating in a single application. It needs to be noted that the infrastructure is not limited to a single system or application and can be used as an infrastructure to build future tools, systems, applications, etc. on which will increase the usability of the infrastructure as this will provide a means to share logic between different systems, any system that the user chooses to use.

Acknowledgements. The author would like to acknowledge Delft University of Technology and Arup for their continuous support and valuable experimentation grounds.

References

1. Aish, R.: Introduction to GenerativeComponents, a parametric and associative design system for architecture, building engineering and digital fabrication. White paper. Bentley systems (2005), http://www.bentley.com (cited May 20, 2007)
2. Coenders, J.L.: NetworkedDesign, next generation infrastructure for computational design. PhD dissertation, Delft University of Technology (2011)
3. Eastman, C.M.: Building Product Models: Computer Environments Supporting Design and Construction. CRC Press, Boca Raton (1999)

Digital Technologies for Evolutionary Construction

Jan Knippers

Abstract. Todays engineering of building structures is based on model thinking. Load bearing structures are developed by choosing solutions from a catalogue of systems (e.g. beam, truss, arch .) established approximately 100 years ago with the criterion of calculability and the methodology of analysis. Digital modelling, analysis, and fabrication allow not only for complex geometries but also a return to redundant, overlaying and indeterminate structural systems, as they were used when the development of load bearing structures was based on a craftsmans experience, and not on verifiable structural analysis. To go one step further, digital technologies allow for a new approach to natural systems, an option, which is neither fully realized nor explored yet.

1 Digital Chain of Design and Fabrication

Since the mid-1990s grid shells such as the roof over the courtyard of the British Museum in London (A.: Foster and Partners, E: Arup; C: Waagner Biro), or the DZ Bank in Berlin (A.: Gehry Partners; E.: Schlaich Bergermann and Partners; C.: Gartner) have become important role models to the building construction industry in terms of developing a digital chain of design and fabrication [2].

While these examples can clearly be identified as shell structures, the roof over the shopping complex My Zeil in Frankfurt (A.: Massimiliano Fuksas; E.: Knippers Helbig; C.: Waagner Biro) show a more complex load bearing behaviour. It neither belongs to the category shell nor to grillage, but shows areas which are predominated by bending moments and require a regular support by columns and others, which show larger axial forces as a shell and allow for greater spans. The transition between the two models shell and grillage is fluent [3].

Jan Knippers
Knippers Helbig Advanced Engineering, Stuttgart, Germany

Free formed structures require a new understanding of engineering and redefine the role of the engineer in the design process. The meshing of the grid, in particular, is a new design step which requires new competences and new tools. In case of the Frankfurt roof it quickly became clear that standardized tessellation or triangulation tools were not generating solutions reflecting the aimed appearance nor integrating structural requirements. The main reason was that the software available at that time was starting the generation process from the edges. Therefore, small flexible software tools have been developed which can easily be adjusted to fit the design direction. Considering the high number of restrictions and edge conditions this development led to a strong design as well as the best suitable structural solution.

Previously in grid shell design, the optimization of the geometry and the construction sequence were the focus of the engineers. The design goal was straightforward: reducing steel usage, i.e. the member sizes, and reducing the number of different construction elements, i.e. member lengths and vertex angles. The intelligent

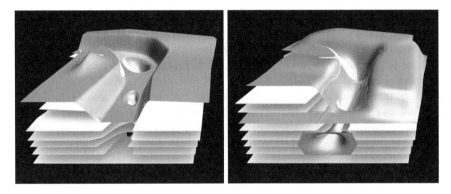

Fig. 1 Initial shape of the architect (left), shape after modification by engineering process (right) (images: Knippers Helbig)

Fig. 2 Initial mesh developed by automated mesh generators (left), final mesh after optimization (right) (images: Knippers Helbig)

Fig. 3 Grid shell MyZeil, Frankfurt, 2008. A.: Massimilano Fuksas, Rome. E.: Knippers Helbig Advanced Engineering Stuttgart. C.: Waagner Biro, Vienna (photo: artur/Barbara Staubach)

solution of a nodal connector was in the focus of the design process and therefore often also exposed as an important aesthetic element.

Today, the situation is often more difficult. Aesthetic understanding tends towards flowing spaces and shapes such as those seen in the Frankfurt roof that deviate from structurally optimized shells. This leads to more and more complex geometries which require a continuous process of engineering from the very first architectural idea to the assembly on site. This process, the digital chain, consists essentially of the following steps:

- Optimizing shape and all-over geometry
- Mapping of load bearing mesh on 3D geometry
- Design and analysis of structure
- Transfer of data to production

Several parties are involved in this process: the architect at the beginning, the contractor at the end and the structural engineer in the centre as a mediator between the aesthetic ambitions of the architect, the technical capabilities of the contractor as well as the financial restrictions of the budget. Therefore, the role of the engineer has changed from the inventor of the efficient structural solution to the manager and administrator of the digital chain.

2 From Structural to Geometric Engineering

Increasingly, it is requested that aspects of sustainability are integrated in the load bearing system. Especially for very large structures such as the Baoan International Airport in Shenzhen (A.: Massimiliano Fuksas; E.: Knippers Helbig), with a total length of more than 1,4 km, the conditions inside and outside the building are not constant for the entire length, but need local adaption.

In the specific case of Baoan International Airport, a modular faade cladding based on the idea of an oversized expanded metal sheet is developed and mapped on a free formed 3D surface. The size and the slope of the glazed openings are adjusted locally to meet, among others, the visual appearance, natural daylight, and solar gain requirements.

Throughout the design phase, the basic shape was changed several times, and the geometry had to be adjusted. Therefore it was of crucial importance to develop a strategy which gave the architect the maximum amount of freedom in developing the aesthetic and functional shape of the building, without delaying the process of detailed design, structural analysis and construction. A data model was developed which allowed the architect and engineers to communicate easily. Based on the 3D surface given by the architect, the entire system including the interior and exterior cladding, as well as the load bearing system is evolved using parametric tools mainly based on Excel and Rhino. The developed data set served as a numeric basis for the parallel check of structural integrity: The structure itself is a simple two layer space frame with a span of 80 m which has to follow the modular arrangement of the cladding.

Fig. 4 Terminal 3 Baoan International Airport Shenzhen, 2013 A.: Massimiliano Fuksas, Rome. E.: Knippers Helbig Advanced Engineering Stuttgart. Local Partner: BIAD, Beijing (rendering: Fuksas))

 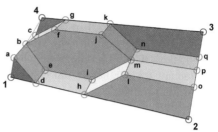

Fig. 5 Physical model of facade (model and photo: Knippers Helbig)

Fig. 6 Definition of facade panel (image: Knippers Helbig)

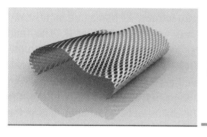 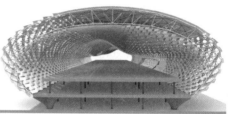

Fig. 7 Parametric model of outer cladding (rendering: Knippers Helbig)

Fig. 8 Parametric model of total roof including space structure and cladding (rendering: Knippers Helbig)

Fig. 9 Site photo June 2011 (photo: Fuksas)

The load-bearing system is looked at as a part of the entire data model, thus the role of the engineer is extended from design and analysis of the load bearing system to handling the geometry of the 150.000 different three-dimensional folded panels and 350.000 members of the space structure.

3 Physical Form-Finding in the Digital Design Process

Examples like the Shenzhen airport demonstrate the current stage of parametric design: usually parametric tools are used to evolve and define the geometry by mathematical rules. The check of structural integrity and the materialization of the structure are parallel or subsequent steps which might necessitate an adjustment of the parametric geometry model. The next step is therefore the integration of a physical form-finding in the digital chain of design and production, which is currently not commonplace in industry. A fitting example is the research pavilion built by the members of the Institute of Computational Design (ICD, Prof Achim Menges) and the Institute of Building Structures and Structural Design (ITKE, Prof Jan Knippers) at the University of Stuttgart in 2010. The process of design and fabrication as well as the structural system are described in detail in [4, 5]. The structure is entirely based on the elastic bending behavior of 6.5mm-thick birch plywood strips. The strips are robotically manufactured as planar elements, and subsequently connected to coupled arch systems as shown in Fig. 3. The final torus shape of the pavilion was a result of the radial arrangement and interconnection of the self-equilibrating arch system. Due to the reduced structural height, the connection points locally weaken the coupled arch system. In order to prevent these local points from reducing the structural capacity of the entire pavilion, the locations of the connection points between the strips need to change along the structure, resulting in 80 different strip patterns constructed from more than 500 geometrically unique parts. The necessary condition to realize this construction was a continuous computer aided design, simulation, and manufacturing process (CAD-CAM).

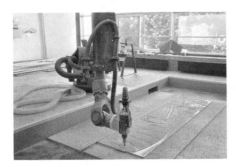
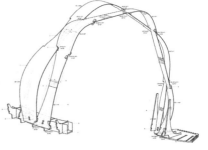

Fig. 10 Cutting of plywood strips, ICD/ITKE research pavilion University of Stuttgart (photo: Oliver David Krieg)

Fig. 11 Model of bent plywood strips (illustration: Wchter, Eisenhardt, Vollrath)

Fig. 12 ICD/ITKE research pavilion, University of Stuttgart, 2010 (photo: Roland Halbe)

4 From Digital Design to Natural Structures

Despite not being directly inspired by a biological model, the final design of the pavilion inevitably reminds of a natural structure. This may be due to the fact that the seemingly innovative principles used are frequently found in nature, but not in building construction today. Often, natural constructions consist of only a few basic components that are geometrically, physically and chemically differentiated. In this sense they are fundamentally different to most architectural constructions [6]. These architectural structures consist of highly differentiated material and functional components (e.g. steel for the structure, glass for the envelope and different plastics for the installations ...), which are themselves geometrically simple and can be assembled by being added to each other to build up the desired construction.

The pavilion, however, follows natural design principles: the homogeneous construction due to the usage of a single type of material only, the parametric differentiation of the plate geometry with a uniform basic typology, and the form definition through large elastic deformations.

The design and construction principles demonstrated by the pavilion might encourage architects and engineers to look more closely at natural structures.

5 Conclusion

Digital design and fabrication technologies have a potential for architecture which is neither fully recognized nor explored yet: They may lead us back from a model based approach using a limited canon of options, to natural structures with a nearly infinite variety evolved by mutation, recombination, and selection. Today, the technical requirements for a transfer of knowledge from nature into construction engineering are much more prevalent than only a few years ago. Therefore, the study of biomimetics for architects and engineers is even more relevant today than ever before.

References

1. Wilkins, D.R.: Getting Started with LaTeX, 2nd edn (1995), http://www.maths.tcd.ie/ dwilkins/LaTeXPrimer/
2. Knippers, J., Helbig, T.: Smooth shapes and stable grids. In: Shell and Spatial Structures: Structural Architecture - Towards the future looking to the past. Proceedings of IASS Symposium 2007, Venice, Italy, pp. S. 207–208 (2007)
3. Knippers, J., Helbig, T.: The Frankfurt Zeil Grid Shell. In: Evolution and Trends in Design, Analysis and Construction of Shell and Spatial Structures, Proceedings of IASS Symposium 2009, Valencia, Spain, pp. S. 328–329 (2009)
4. Menges, A., Schleicher, S., Fleischmann, M.: Research Pavilion ICD/ITKE. In: Proceedings of the FABRRICATE Conference 2011, Stuttgart, London (2010)
5. Lienhard, J., Schleicher, S., Knippers, J.: Bending-active Structures - Research Pavilion ICD/ITKE. In: Taller, Longer, Lighter - Meeting Growing Demand with Limited Resources, Proceedings of IABSE-IASS Symposium 2011, London (2011)
6. Knippers, J.: Speck, T.: Design and Construction Principles in Nature and Architecture. Submitted to: Bioinspiration & Biomimetics

Combinatorial Architecture

Enrique Sobejano

1 Combinatorial Architecture

Architects have always tried -using the means available to them- to find strategies for mastering the formal, geometric and constructive structures that take part in the design process, by fixing a limited number of elements and defining the laws that connect them to one another: a combinatorial process that, for centuries, was essentially based on drawing and models. These traditional tools made it possible to maintain a specific balance between the idea, the hand that drew it, the scale of representation and the instructions given to the builder; or in other words, between the design and the spatial and constructive execution of the building. However, over the past several years, manual drawing has been inevitably substituted for computer-assisted design, which for both technical and economic reasons has since extended almost universally. What began as just a new tool for drawing, is now substantially modifying the historic connections between the design and the architectural production, to such an extent that some people believe that we are immersed in a digital revolution that not only affects the manner of representation, but also the formation of creative processes that the designer would be incapable of devising without their assistance.

Upon observing the growing interest of digital architectures advocates in defending the importance of the process over the result, rejecting the alleged rigidity of Euclidean geometry in favor of topological dynamism and valuing the computer-based simulation of formal evolutionary processes over the constructions themselves, one might ask whether this apparent indifference to the material nature of architecture isnt hiding a latent conflict between the potentiality of forms generated digitally and their construction. Process/diagram/topological transformation are concepts that appear to be in opposition to those -form/type/tectonics- that had dominated theoretic and academic discussion before the arrival of computationally informed architectural design.

Enrique Sobejano
University of the Arts, Berlin, Germany

As it tends to occur when a technological innovation emerges -which is clearly the case of digitalization in the field of architecture- radically opposite positions initially form. What some people had greeted with enthusiasm and unlimited confidence in its revolutionary possibilities, others looked upon with disinterest, if not disdain, as simply a new tool for geometric drawing and representation that would have little real impact on the creation and development of architecture. Now that enough time has passed since the topic of digitalization was introduced into the architectural debate during the 1970s and 80s, it might finally be possible to form a more objective reflection conscious of the fact that architecture is, above all else, the result of a direct link between ideas and their materialization.

While building takes time, digitally generated architecture is most frequently characterized by a fact: its immediacy, the speed of its production, and therefore for its ability to be considered as a final product in and of itself. This likewise fuels the fantasy that the purpose of architecture lies in the new methods of digital modelling, which in many cases, leads to a progressive distancing between a graphic representation and its constructive reality. Some architects are thereby turned into the creators of virtual dreams that inundate publications, university projects and architectural competitions; inspired in digital utopias that only appear to be linked with the other social, mechanical or urban utopias that had arisen under 20th century ideologies. The metaphorical dimension inherent to every architectural proposal -that we recognize, for example, in Le Corbusiers attraction to machinery or the prophetic technical designs of Buckminster Fuller- has been replaced in the imaginations of architects of the digital generation by genetic, evolutionary, topological and fractal models, etc. that are often adopted with an acritical faith that perhaps forgets the fact that architecture is not a scientific activity and therefore its paradigms cannot be extrapolated literally.

The mechanist idea of the architectural avant-garde of the 1920s, metaphorically represented by large steamers, airplanes and power plants, was interpreted by the futurists, constructivists and architects of the Bauhaus, in buildings like infrastructures, houses like machines for living and cathedrals of industry. However, despite the literal nature of many of their references, these modern pioneers used the technical and material innovations of their time as a basis for their formal proposals in such a way that they never totally dissociated from their constructive dimension. On the other hand, the visionary architectural and urban representations of the 1960s looked for references in the mechanism to explain an illusion, in other words in the effect more than the object. They were utopias that had the common denominator of a fascination with movement, with industrial dynamism and for urban organisms in constant transformation: metaphors in this case for actions referring to a future world that its authors would never build. When Archigram, Yona Friedman or Constant proposed visions of moving buildings, continually growing cities or huge urban megastructures, the necessary technical capabilities for their design and construction didnt exist, and so they were relegated to publications and the collective memory of unattainable utopias. They established a framework that defined an imaginary world, populated by technological artifacts, based in an acritical acceptance of all things modern, a vision of a future nurtured by technology and from science

Fig. 1 Archigram, Plug-in City

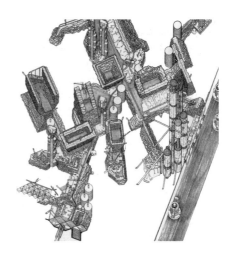

fiction, translated into an unlimited confidence in progress. Part of the romanticism that we remember them by today is based on proposals that were never built, such as the transformable assemblies of *Plug-in City*, the inhabitated megastructures of *Ville Spatiale* or the awesome urban spaces of *New Babylon*. When utopias transform into actual buildings, the poetic sentiment of the unattainable fades away. This is also happening now, at a time when some of the formal digital utopias have begun to materialize. In the previous examples, the tools used during the design and development process were still manual drawing, collage, models or photography. Consequently, the breach that could have emerged between representation and reality resided in the capability or incapability of the existing technologies to translate a two-dimensional image or a model into a building. Today, however, the technological innovation is being produced in the tool itself -software and computers- with much greater intensity than in the techniques, materials and building systems, and in such a way that the new digital utopia is in danger -which is, in fact, currently taking place- of distancing itself from the constructive purpose of architecture, and therefore the essential link between the world of ideas and their materialization. Therefore, the transfer of digital designs into building produces a gap where the lightness and dynamism transmitted by the virtual images often contrasts with the gravity and materiality that they display upon being built. The issue is therefore not whether or not digital methods are accepted, but rather to what extent the alleged computer revolution deepens the constructive purpose of architecture.

The mechanist utopias of the last century have been replaced in our times by biological processes and their architectural translation by way of computer design programs. The project is no longer interpreted starting from objects, typologies or mechanisms, but rather as the result of combinatory variations generated by algorithms, topological geometries and open systems that are applied as much to the volumetric and spatial definition of buildings, as to the transformation of urban landscapes or ecologically sustainable proposals. It is argued that, like the natural evolutionary processes, the successive mutations and adaptations -with allusion to the seductive biological metaphor of the balance between *chance* and *necessity*, as stated by Jacques Monod- should eventually lead to the most appropriate architectural design. In reality, the work of the architect, in his search for solutions to variable parameters -structure, cost, function, climate, natural light, etc.-, has always consisted of defining the laws that connect them to one another and the instructions that allow for the mediation between an idea and its construction. The parametric and algorithmic design that is enthusiastically defended by many today, could also be interpreted as a new manifestation of the existing permanent dichotomy between the uniqueness of a singular piece of work and the other multiple formal probabilities that might have been adopted. During the 1980s, this duality was explained using the notion of type, understood to be the formal internal structure shared by determined buildings. This non-utopian typological thinking looked for its roots in the past, not the future. It was the history of architecture itself that would have fixed the types as distillations of centuries of constructive evolution. While the typological interpretation also incorporated -as observed by Rafael Moneo [1] - the idea of change and transformation, it was understood that the architect should first begin by identifying a type that would later be destroyed or modified. The transition from the structuralist line of thought that pervaded these conjectures towards the post-structuralist model that was essentially inspired by the work of Gilles Deleuze -habitually cited as an theoretic argument for architectural digitalization- was translated into the rejection of the notion of typology in favor of a idea of architecture in dynamic and organic terms. The typological series are now being transformed into evolutionary processes, generated from abstract diagrams that contain, much like DNA, the multiple possible evolutionary combinations of the architectural form that have been made popular today by new software programs. The relationship between the genetic algorithm and architecture proposed by Manuel de Landa following the Deleuzan line of thought [2], therefore opens a field of possibilities to the designer in which the focus is no longer on the final representation -the platonic ideal based in Euclidean geometry- but rather on the morpho-genetic potential, now referred to the topological geometry, and to the contemporary conception of the evolution of living organisms. The rigidity inherent to the platonic conception of type is replaced by the evolutionary metaphor, so that a specific design is nothing more than an intermediate point, a random event in a combinatorial system: the generating rules are no longer based on components, but rather on analogous processes found in nature.

The fascination wielded by natural metaphors in architecture is evidently nothing new: from the intimate link of primitive cultures to topography and landscape, to the imitations of plant forms that have been used for centuries as ornamentation and the organicist experiences of the last century; nature exerts a permanent seduction on architects. The direct correspondence of natural forms with the diagram of the forces that acted on them during their formation process -as was proposed almost a century ago by the biologist DArcy W. Thompson [3] is, for example, a model that is commonly referred to today in the field of digital architecture, despite the fact that its original scientific bases are now obsolete. But (we) architects have never really been completely interested in scientific truths, rather in their metaphoric

Combinatorial Architecture 59

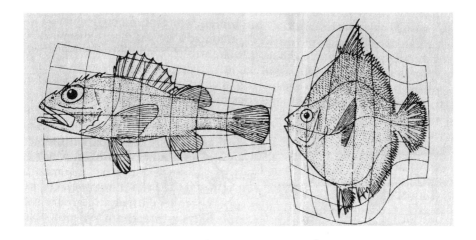

Fig. 2 D'Arcy Thompson, On Growth and Form

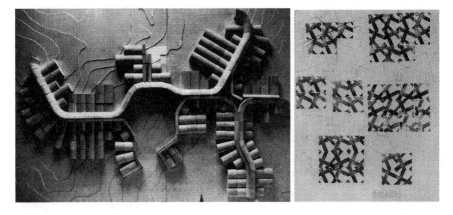

Fig. 3 *(left)* Jorn Utzon, Farum Centre   **Fig. 4** *(right)* M.C.Escher, Drawing

interpretation, that which architecture can simulate using them as a reference. When we admire the movement diagrams of Louis Kahn for Philadelphia or the organic additive models of John Utzon, we are not literally evaluating their scientific veracity -nor did probably their authors- but rather their ability to simulate natural processes and their application in the field of architecture. As Slavoj Zizek observed, virtual reality doesnt imitate reality, but rather simulates it [4], in the same way that new digital technologies dont merely imitate nature, but rather make the underlying combinatorial mechanism that they generate visible.

The combinatorial processes that we recognize, for example, in the fugues of J.S. Bach, narrations of Georges Perec, drawings of M.C. Escher, ornamental geometries of Islamic art or in the formal structures of nature -all of these brilliant precedents

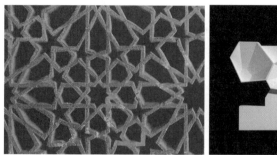
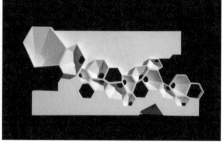

Fig. 5 *(left)* Alhambra Granada, Detail

Fig. 6 *(right)* Nieto Sobejano, Cordoba Art Centre

Fig. 7 Nieto Sobejano, Temporary Market Madrid

and references to current digitalization- surprisingly express the simplicity that is hidden behind the complexity of every musical, pictorial, literary or architectural creation. As with other creative endeavors, I am also conscious of the fact that the process by which an abstract idea is eventually transformed into a concrete result -the architectural work- is constantly nourished by images hidden in our memory.

This subjective and poetic background that all projects entail therefore enters into conflict with the mathematical exactitude to which digitally developed projects aspire. It is therefore difficult for me to accept that the only and inevitable path in the digital Zeitgeist lies in the direct transcription of diagrams or algorithmic and parametric processes. On the contrary, I believe that architecture is always the result of an interpretation of multiple and apparently unconnected circumstances that strangely begin to look like one another; that only through building it is possible to select one of the multiple formal, geometric and constructive combinations that relate to a place and its memory. After all, aren't *form*, *geometry* and *construction* not just different manifestations of a same combinatorial concept?

References

1. Moneo, R.: On Typology, Oppositions, vol. 13. MIT Press, Cambridge (1978)
2. De Landa, M.: Deleuze and the Use of the Genetic Algorithm in Architecture (2001)
3. Thompson, D.: On Growth and Form. Cambridge University Press, New York (1945)
4. Zizek, S.: Lacrimae rerum. El ciberespacio o la suspensión de la autoridad. Random House Mondadori, Barcelona (2006)

Codes in the Clouds
Observing New Design Strategies

Liss C. Werner

1 Introduction

Looking at architecture through the lens of currently emerging, advanced architectural design tools, strategies and techniques in the 21st century are as questionable as looking through the lens of a 2-dimensional Rapidograph drawing. Architecture is neither determined by complex geometries digitally established through scripting, nor by the one dimensional perception through a hard line drawing. The role of architecture is currently being challenged through almost overwhelming advanced digital techniques merged with traditional strategies in order to find its new identity within a post-mechanical towards a biological computing age of construction. We are seemingly fooled by the understanding that we do conquer the boundaries of geometry and form production. The paper discusses the connection point of advanced digital architecture as tool to complement and bring forward professionally and culturally established parameters, such as material tectonics, perception or architectural pattern. If we can establish an approach to architecture, that includes our current tools, we are able to gain an unexpected understanding of existing architectural knowledge and at the same time open up new opportunities through applying emerging insights from other fields and disciplines to architectural thinking. "An architecture machine that could observe existing environments in the real world and design behaviours from the parent would furnish the architect with both unsolicited knowledge and unsolicited problems." [1] Complexity in architecture now means a critical reflection of digital architecture within the existing cultural and material context. A key to crossing the boundaries towards contemporary development of a novel architectural understanding may be found in research results of cognitive sciences starting from Herman von Helmholtz. The paper investigates a systemic approach based on biological computing and cybernetics in order to understand the

---

Liss C. Werner
Dessau Institute of Architecture, Dessau, Germany
Humboldt University, Berlin, Germany

opportunities of digital parametric design processes and tools. Discussing the implementation of culture, tradition and hard facts of human neural processing of the environment it reflects on the beginnings of evolutionary architecture and also critically observes and challenges the current usage of scripting and generative design towards a holistic approach for current design challenges as design strategy.

2 Outline

The first section of the paper, *Learning from Helmholtz*, will observe the beginnings of combining neuro-physiology, measuring devices and proportions and their benefits for todays parametric design tools. Criteria are offered which open opportunities to support our current design thinking beyond the understanding of complex geometry and software packages. The recently published book Cognitive Architecture edited by Deborah Hauptmann and Warren Neidich (Hauptmann and Neidich 2010) may be seen as a kind of manual following the research results by Hermann von Helmholtz and Jean Piaget in the 19th century. Cognitive Architecture brings together a number of relevant essays from esthetics, neuro-sciences, politics and architecture to suggest a reconfiguration of the still existing understanding of form-giving towards an architectural design thinking of networks and data flow.

The second section, *Beyond Parametricism*, contextualizes Parametricism within a 2nd order cybernetic environment as a constructed system rather than a style. The model of parametric architecture as suggested by Patrick Schumacher (Schumacher 2009) is revisited whereby attention is given to an approach including human behavior and human learning. Culture specific issues such as pattern, ornament, tectonics, personal distance and complexity of syntax in language may be taken in consideration when designing future tools for generating architecture. Thus the third and final section, Coding Strategies, finally suggests a multi-dimensional, complex and flexible design model for spatial-differentiation-logic adjustable to context and environment opening a new territory for architectural profession and research. When understanding architectural results as effects caused by continuously adapting human brain activities and transformation of culture they can be regarded as extensions of the designer. Following the principle of identifying the tools with which a product is made the evolution of design strategies unfolds into a hybrid of *multipolar behaving Neurotecture*.

3 Learning from Helmholtz

Design strategies are constructed of tools and decision-making derived from the technology used and the designers or architects skills. They are also informed by personal experience, visual / cognitive perception and physical interaction in space; meaning that they are coded by epistemological and physical data. *Hermann von Helmholtz* (1821-1894), a German neurophysiologist during the $19^t h$ century, can be regarded as the first scientist studying *biological computation* from a neuroscientists point of view in conjunction with physiology and esthetics. His research of how to measure the speed of nerve signals, using a nerve of a frog and a muscle of a calf as well as a galvanometer as timing device Helmholtz provided groundbreaking information for early neuro-sciences. Helmholtz studied the mathematics of the eye and established ideas on the visual perception of space which later became the foundations for his publication *Die Lehre von den Tonempfindungen als physiologische Grundlage fr die Musik*, On the Sensations of Tone as a Physiological Basis for the Theory of Music (Helmholtz 1877). His view from his first publication in 1877 towards harmony and general esthetics, which are apparent in music and architecture alike might describes an avenue to formulate an approach to future design strategies including the modus operandus of the human body: Hence I feel that I should not be justified in passing over these considerations, more especially as they are closely connected with the theory of sensual perception, and hence with physiology in general. [2]

The model of input and output as form and form of communication within a network has been looked at by his successors, the cyberneticians *Norbert Wiener* (mathematician), in the 1940s, *Ludwig von Bertalanffy* (biologist) and *Warren McCulloch* (neurophysiologist at MIT) working on artificial intelligence in the 1950s (Pask, 1982).Within the last decades of the 20th century Gordon Pask, John and Julia Frazer together with their students at the AA applied the results of Data Structures as a *Generative Toolbox* to architectural models such as *Intelligent Beermats* (John, Julia and Peter Frazer, 1980) or *Experimental Neural Network*, (Miles Dobson, student , unit 11, 1991) establishing computing without computers [3] and later using state of the art software at that time. Gordon Pask was far ahead of his time and saw the future of architectural practice not so much to design a building or city as to catalyse them: to act that they may evolve. [4]. Pasks ideas show clear relevance for todays parametric design approach (Frazer 2007). Within the realm of architecture artificial intelligence is becoming increasingly popular and necessary to create high performative buildings and cities.

Nevertheless observing digital design strategies since the 1970s the issue of ordered beauty and esthetics within this field has been addressed only just recently. Aesthetically, it is the elegance of ordered complexity and the sense of seamless fluidity, akin to natural systems that constitute the hallmark of parametricism. [5]
Parametricism is the construction of a system with interdependent nodes and attractors, these techniques are highly efficient for remodeling forms(Leach) in many cases leading to an esthetically pleasing result, where individual parts of the architecture act in the background, as meta-system, providing the rules for smooth surface modulation. In regards to the construction of beauty in arts Helmholtz takes a similarly systemic point of view in saying The more we succeed in making the harmony and beauty of all its peculiarities clear and distinct, the richer we find it, and we even regard as the principal characteristic () that deeper thought, reiterated observation and continued reflection shew us more and more clearly the reasonableness of all its individual parts. [6]

4 Beyond Parametricism

Form is a verb and not a noun. Patrick Schumachers approach to parametricism as a style is to be observed highly critical when going beyond the reduction of pure geometrical or number based data, location models or site-embedded systems. Even the consideration of advanced BIM models, mainly to develop ecologically and economically well performing buildings and cities, demanding higher differentiation within the design process, outcome and decision-making, the previously mentioned inevitably important points for creating architecture are missing. Parametric design depends greatly on a manipulation of form that might appear visually interesting, but is often superficial. [7]

Freeing *Parametricism* from its brand as a style and opening it up to learning from a rather traditional approach of vernacularity, phenomenology as discussed by Christian Norberg-Schulz or even act as an extension of Kenneth Framptons theory of *Critical Regionalism*, would, in combination with knowledge based on Helmholtzs research and current findings in neuro-sciences offers an enormous amount of flesh to formulate a contemporary realm and understanding for architectural theory and practice in the 21st century. Based on existing knowledge architecture can evolve. Christian Norberg-Schulz for instance describes The functional approach therefore left out the place as a concrete *here* having its particular Identity [8] and place-making as Being qualitative totalities of a complex nature, places cannot be described by means of analytic, *scientific* concepts. [9] Functional approach in the realm of parametric design as applied so far refers partly to the program but mainly to the hard performance of the building or city, respectively the embedding of human movement (Space Syntax) and still leaves the gap of culture, poetics and identity. The question here is, if architecture actually wants to overcome the idea of discarding style. At the same time we do have to ask the question if style is not actually necessary for formulating cultural identity.

One of the few project that did include culture within the design process and design outcome is Joean Nouvels Louvre Abu Dhabi on Saadiyat Island. The project pushes digital design and manufacturing tools and strategies to their limits by considering an extreme climatic environment, but also cultural aspects within form and surface topography. The design for the 183m dome is driven by an intricate weaving and interlacing of traditional Arabian pattern, digitally manufactured. Thus generating form through a cultural parameter and advanced manufacturing techniques.

In his article *The Autopoiesis of Architecture*, published in 2002, Schumacher refers to Luhmann's theory of modern society as a functionally differentiated society is embedded in his general theory of social systems. [10], which builds upon the concept of autopoiesis, a term coined by the biologists *Humberto Maturana* and *Francisco Varela* in 1972. Maturana on the other hand states, that an autopoietic machine is a machine organized, defined as a unity, as a network of processes of production (transformation and destruction) of components which: (i) through their interactions and transformations continuously regenerate and realize the network of processes that produced them; and (ii) constitute it (the machine) as a concrete unity in space in which they (the components) exist by specifying the topological domain of its realization as such a network. [11] (...) the space defined by an autopoietic system is self-contained and cannot be described by using dimensions that define another space. When we refer to our interactions with a concrete autopoietic system, however, we project this system on the space of our manipulations and make a description of this projection. [12]

Autopoiesis as described by Maturana and Varela refers to closed systems, which in light of globalization might not be applicable anymore. As architects designing more global public than local private spaces we may want to open the definition of autopoiesis as closed system to a definition of *open system, dissipative structures* incorporating the dynamics of non-equilibrium systems [13] accepting and being shaped by perturbations, disturbances or in the widest sense, influences. Other than Maturana who would be reclined to speak of self-organising systems when speaking of autopoietic systems, Pirgogine only regards systems of self-organisation in a highly dynamic environment. Independent of size or scale the challenge of how to carry parametricism forward remains.

The above arguments on one hand describe a sociological and biological inclusive discourse, while on the other hand informing development, usage and application of future parametric design strategies. If architecture works as an agency, a sub-system, it does have an environment, which I may call culture. Schumachers statement that There is a global convergence in recent avant-garde architecture that justifies its designation as a new style: parametricism. [14] is hardly coherent with his model of autopoiesis being biological life-processes as the circular self-reproduction of recursive processes [15] for architecture, but offers a challenge and opportunity at the same time.

Thus a truly parametric architectural approach may want to refer to akin researchers of autopoiesis, *radical constructivism* (Ernst von Glasersfeld 1995) or later epistemology. So Schumacher has touched upon the integration of culture and inter-cultural knowledge as well as culture-specific behavior, but not integrated into *Parametricism* yet. The opportunity for architecture to embed culture, language and knowledge into parametric architectural design strategies is already suggested in Schumachers essay Parametricism A New Global Style for Architecture and Urban Design. [16]

5 Coding Strategies

Coding Strategies describes the last section of this paper. It acts as a conclusion and incubator at the same time. Following the theories of *Warren McCulloch* , MIT, architecture may include considering subsystems, urban or local as living systems or informative structures for biological computing. Referring back to Hermann von Helmholtz, Gordon Pask around 100 years later states in Micro Man Computers and the Evolution of Consciousness that the approach to simulating the human brain activity was largely superseded by attempts to model behavior instead. [17] I do want to dwell on this for a while and refer to Mark Weiser, who was one of the first to envision computers as a pervasive part of everyday life, the intelligent agent was the metaphor for the computer of the future. [18] And also to Jordan Crandall in his essay Movement, Agency and Sensing: A Performative Theory of the Event. According to Crandall, Data-intensive, multi-agential environments characterized by combinations of inexpensive sensors, interoperable clusters of computing platforms, high-bandwidth networks, and large-scale coordination among different database systems are accompanied by analytical tools, modes of inquiry, forms of being, and practice of movement. [19] Jordan Crandall, artist and media theorist, actually manages to at least theoretically combine the factors that might be incorporated in parametricism beyond parametricism to establish coding strategies. He looks at the transient user of architecture and in his essay refers to Katherine Hayles who in her own right investigates into distributed cognition occurring on creating interrelated systems among sub-cognizers, readers, and relational database. Crandall further states that the human body itself can already be understood as a distributed system the physiology through which thought occurs composed by networks of actors operating at multiple scales, and which extends into the surrounding world. [20]

Taking in consideration the research by Hermann von Helmholtz who incorporated the human body, its neural network, physiology and related understanding of space into comprehension and description of the world around us I do question the reduction of parametricism to geometry and hard data and offer a development of design strategies accordingly. Parametric design tools, especially software packages such as *Processing*, *GHowl*, *CSharp* and possibly *Grasshopper* are powerful tools, which we may be able to extend in including non-geometrical but cultural parameters.

The relevance of biological computing as suggested by Philip Beesley in his project Hylozoic Ground [21], exhibited at the Venice Biennale 2010, may also be taken into account. Beesley strongly follows Gordon Pasks first interactive computationally inspired architectures such as *Musicolour* (1953) or *The Colloquy of Mobiles* (ICA, 1968), both describing human machine learning. Beesley as Grdon Pask did looking towards an interactive computed architectural environment. In contrast to Gordon Pask, who investigated in biological computing Philip Beesley is working on physical computation influenced by biological behavior. Using its tendrils, fronds and bladders to lure visitors into its seemingly fragile web of laser-cut acrylic matrices, this work blurs the distinction between organism and environment. [22] Except from aiming at the actual application in real-manufactured architecture the focus of the paper locates itself within the shifting realm of established architectural theory, still remaining within post-modernity and the limitation to typologies. It hints towards understanding the current human condition, prosthetics as inclusion of computing abilities and an understanding which on one hand might be more complex, but actually closer to our native human make up and suggests a design strategy described as NEUROTECTURE.

References

1. Negroponte, N.: The Architecture Machine: Towards a More Human Environment. MIT Press, Cambridge (1970)
2. Helmholtz, H.: On the sensations of Tone as a Physiological Basis for the Theory of Music, London (1895)
3. Frazer, J.: An Evolutionary Architecture (1995)
4. Ibid
5. Schumacher, P.: Parametricism A new Global Style for Architecture and Urban Design. In: Digital Cities. Wiley, London (2009)
6. Helmholtz (1895)
7. Leach, N.: Parametrics (2009)
8. Norberg-Schulz, C.: The Phenomenon of Place. In: Nesbitt, K. (ed.) Theorizing a new Agenda for Architecture, 1st edn., Princeton Architectural Press, New York (1996)
9. Ibid
10. Schumacher, P.: The Autopoiesis of Architecture. In: Latent Utopias. Springer, New York (2002)
11. Maturana, H.R., Varela, F.J.: Autopoiesis and Cognition. D. Reidel Publishing Company, Dordrecht (1980)
12. Ibid
13. Prigogine, I.: Self-Organization in Non-Equilibrium Systems. John Wiley & Sons Inc., New York (1977)
14. Schumacher (2009)
15. Schumacher (2002)
16. Leach (2009)
17. Pask, G.: Microman. Macmillan Publishin Co., Inc., New York (1982)
18. Weiser, M.: The Computer for the 21st Century, Sci. AM 94 (1991)
19. Crandall, J.: Movement, Agency and Sensing: A Performative Theory of the Event. In: Hauptmann, D., Neidich, W. (eds.) Cognitive Architecture, 1st edn., 010 Publishers, Rotterdam (2011)
20. Ibid
21. Beesley, P.: In: Ohrstedt, P., Hayes, I. (eds.) Hylozoic Ground, 1st edn., Riverside Architectural Press, London (2010)
22. Beesley, P.: Hylozoic Soil. Leonardo, vol. 42. MIT Press, Cambridge (2009)

Integration of Behaviour-Based Computational and Physical Models
Design Computation and Materialisation of Morphologically Complex Tension-Active Systems

Sean Ahlquist and Achim Menges

Abstract. The use of Particle Systems to replicate physical form-finding methods has been well documented; initiated, arguably, with Axel Kilian's CADenary software mimicking Gaudi's hanging-chain methods. The evolution of these algorithms for simulating fundamental physics has seen models formed via forces such as gravity, tension, compression, magnetism, and pressure. The approach seeks geometry organization, not through discreet rules, but through emergent behaviour of interacting forces forming an equilibrium state. The value, though often unseen, is that these resulting models hold information, not only of geometric description, but of dynamic force-active properties, critical for materialisation as well as the understanding and design of the entirety of the interrelation system. This paper shall focus on the development of a Particle System based computational methodology which integrates various critical aspects of force, materiality, and form in the generation of morphologically complex tension-active systems.

1 Introduction

Behaviour-based computational form-finding is structured by three primary functions: simulation of force, modelling of topology, and translation from geometry to materialization (Ahlquist and Menges 2010). With each aspect open for design consideration, the design environment, while constrained, allows for a high degree of flexibility in specifying material descriptions from and within the generated force diagrams. The physics-based nature of a Particle System, defining the behaviour of material and force, is particularly relevant to tension-active systems in their complex surface geometries which are mathematically inexpressible. This paper shall describe, through a series of case studies, these aspects of a behaviour-based design framework exposing the avenues for design articulation and

---

Sean Ahlquist · Achim Menges
Institute for Computational Design, University of Stuttgart, Germany

construction of highly complex and articulated tension-active systems. Defined by textile surface and linear cable networks, the structural rigidity of a tension-active system is realized through a constant distribution of tensile forces. A stable system utilizes an equalized distribution of force, defined as pre-stress, through doubly-curved geometries.

Computationally, tension-active systems are formed via vectorized force elements recognizing fundamental material rationales, in their internal capacity and overall assembly logic. As structurally defined surfaces, precision and exaction of the material properties, assembly constraints, and stability against external forces is vital.(Kilian 2004) However, this is something that cannot be directly coordinated within the simple parameters or solvers that drive a Particle System. The research and projects described in this paper will elaborate upon a strategy that, despite the indirect relation between computational behaviour and material behaviour, provides for calculated design opportunities through a computational form-finding environment and translational algorithms which harvest the full dataset, of geometry and force behaviour.

2 Force Topology

Particles and springs are the primary components of a Particle System that are utilized in this research. A particle provides a position and mass, while a spring defines a force and a vector. Force is determined following Hooke's Law of Elasticity, where the value, simply stated, is calculated by the degree of displacement in length of a spring's equilibrium state (or rest length) multiplied by a spring constant (or strength) (Gordon 2006). It allows for both compression force (actual length *less* than equilibrium length) and tension force (actual length *more* than equilibrium length) to be activated. The most common method for calculating the force and geometry of tension-active systems is in the use of Finite Element Analysis, calculating form directly from specific material properties and defined component assemblies (Moncrieff 2005). Similarly, both methods operate in an iterative manner, applying incremental tension forces to eventually resolve into equilibrium. This is a necessary step as it exposes one of the fundamental challenges in computing tension-active forms: their geometries cannot be described mathematically. The advantage, in comparison, of a Particle System is its ability to rapidly compute the negotiation and iteration of physical forces and to stably realize an equilibrium state through large geometric displacements (Kilian and Ochsendorf 2005).

The utilization of springs provides an avenue to which force behaviour can be investigated. Behaviour is defined as the causal description of transformations occurring via physical laws of environment and material, as well as affects transferred from other behaviours (Goel et al. 2009).The causal relationship in this case works on two levels: the interaction between individual spring elements and the repercussions of the network topology of force elements to the geometry which is realized when forces are resolved into equilibrium. When activating both the local spring mesh topology and the global network as variables, the control over the design of the system becomes quite extensible.

The projects depicted in this paper significantly expand previous research of a programmed generative modelling environment in Processing which uses a cylindrical organization of particles and springs as a base topology for the overall system (Ahlquist and Fleischmann 2009). On the local level, the topology, or *meshType*, varies from highly organized to random. Manipulating certain topologies on the local level allows for basic shape definitions, or *topoTypes*, to be significantly altered. For instance, multiple cylinders can be connected to generate a torus, or the continuity of a cylinder can be broken to generate a plane (Fleischmann et al. 2009). This variability provides a unique opportunity in the study of tension-active systems. All tension-active geometries can be derived from a catalog of three forms: saddle, cone/cylinder, and ridge-valley (Knippers et al. 2010). What offers unique and progressively complex morphologies is the hybridization of these fundamental form conditions. The structure of this computational platform readily allows access to the two design channels of local and global topology in manipulating both the boundary conditions and internal make up of the form-force diagrams, while processing and constraining the dataset to recognize critical material characteristics. (Fig 2.1).

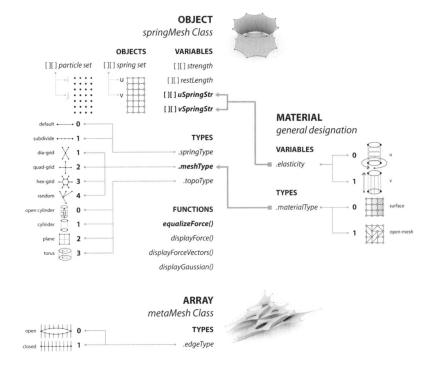

Fig. 2.1 Behavior-based computational structure denoting certain material designations

3 Materialization

The structural dynamics of any architectural system can only be known once its material characteristics have been prescribed. This is of utmost importance when the geometry is a repercussion of the relation of material to structural dynamics. This is the case in tension-active systems: how the material carries the tension defines its geometry. Such is the need for any form-finding method to integrate certain material parameters into the process of form generation. The inclusion of material information is in relation to the relative distribution of forces – how equalized are the forces and how well can the material handle the ranges of distribution – and also in relation to the panelization of a doubly-curved geometry – how can sheets of flat material be generated to act accordingly when stressed to regenerate the complex morphologies defined computationally.

3.1 Relation of Springs to Material

What allows for the rapid computation and large geometric displacements with Particle Systems can be traced through Hooke's Law of Elasticity: a spring's force behaviour works in a linear and infinite fashion. This provides ultimate robustness for force equilibrium to be reached. Unfortunately, this limits the ability to recognize the non-linear behaviour of material. This is not the primary concern though. The most problematic aspect of this is the spring's tendency to exceedingly stretch at the boundary conditions. The resulting geometry has a surface *distortion* at these moments of exceedingly high stresses where the springs are significantly extended, and subsequently results in areas of minimal double curvature (Ahlquist and A. Menges 2010).

An *equalizing* algorithm has been programmed to overcome these moments of minimal double curvature by redistributing the strength values of all the springs thus shifting the force distribution. This introduces two critical aspects. One is that geometry is a primary consideration in how and why the springs' values are manipulated – so that proper doubly-curved surfaces can be achieved. The other is that the force diagram is a relative representation and does not relate to the stiffness of a material.

3.2 Precision in Translation

Tension-active systems can be defined as surface structures where force moves equally in all direction or as mesh structures, often called cable nets, where force is resolved at each locally fixed node. This distinction becomes most apparent in how the geometry and force diagram is translated to information for materialization. For a mesh, each element can be addressed individual. For a surface, which is the concentration in the paper, force and geometry have to be resolved as *fields*.

Conventional methods segregate geometry from force. Panelization of the overall geometry is done via geodesic lines from boundary to boundary and in a manner where each region is primarily singly-curved, the double curvature is mostly neglected. A *compensation* step shrinks the flattened geometry to consider how prestress will affect it when assembled into the complete system (Fig. 3.1a) (Linhard 2009). With the approach defined in this paper, both geometry and force are assessed simultaneously in defining panelization, removing the limitations imposed by the *flattening* procedure. When calculating material elasticity in concert with the spring-force diagram, it can be determined in a generative fashion how much curvature can be accomplished with the material and where the panel boundaries must be defined (Fig. 3.1b). Integrating materialization in such a way allows for panelization, a primary concern in materializing tension-active systems, to be investigated and instrumentalized.

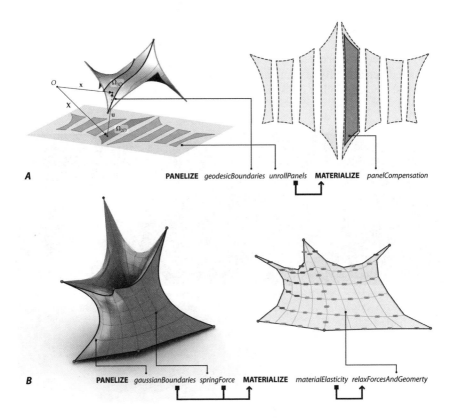

Fig. 3.1 Comparison of *(A)* conventional panelization method by use of autonomous geometry and prestress compensation functions, and *(B)* integrated spring-based method by use of Gaussian curvature, spring force and material elasticity

4 Complex Tension-Active Morphologies

The following studies showcase the particular aspects of topology development and integrated materialization within the Particle System based process. Both focus on a cumulative cellular process, with the outcomes of continuous interconnected surfaces following the basic nature of tension-active systems. The object-based computational strategy collects and interconnects elements at multiple scales (local, global, and meta topologies). Each project slightly shifts the variables at each of these levels to produce significantly different outcomes.

4.1 Surface Articulation

This project exercises that ability to aggregate a series of cylindrical elements in a non-orthogonal manner. While each cell maintains its routine *meshType* and *topoType* parameters, there are local manipulations in the topology to associate boundaries conditions and internal surface conditions in various ways. A standard tension-active form occupies a cubic space, allowing it to have very regular doubly-curved surfaces. By accumulating the global topology in a non-regular manner, it tests the possibility of skewing the surfaces while still achieving anticlastic geometry.

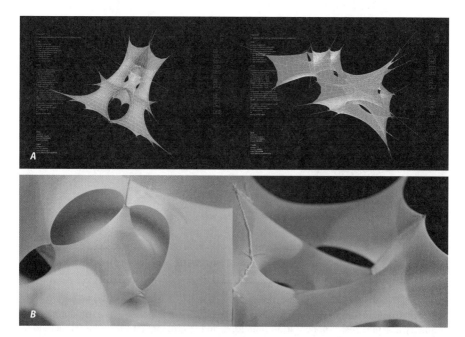

Fig. 4.1 Complex Surface Morphologies Prototype. Yilmaz Demir, Louis Saint-Germain, ICD University of Stuttgart – Sean Ahlquist, Prof. Achim Menges, 2011. *(A)* Computational form-finding model in Processing. *(B)* Large-scale prototype.

4.2 Cellular Articulation

To generate, compute and realize morphological complexity with tension-active systems, knowledge and prototypes (of physical and computational structures) have to move through a progression of development and calibration from simple procedures to complex arrangements. This project exemplifies such development from simple cylindrical assemblies to a highly complex multi-toroidal system. The system is develop through controlling the local level of individual cellular topologies and generating the global level of a meta-structure of associated toroidal cells. The physical prototype showcases the calibration of material translation as a fully tensioned highly articulated series of interconnected anticlastic surfaces. The calibration extends into the inclusion of a primary tensioned mesh (cable-net) to dicate the flow and continuity of tension through the entire system.

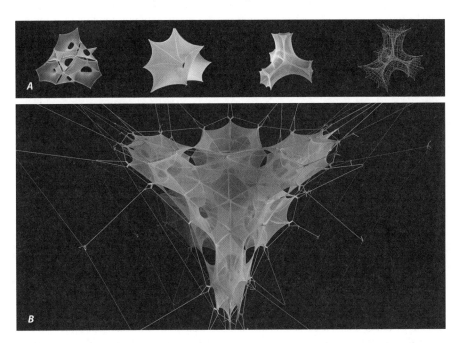

Fig. 4.2 *Complex Cellular Morphologies Prototype.* Boyan Mihaylov, Viktoriya Nikolova, ICD University of Stuttgart – Sean Ahlquist, Prof. Achim Menges. 2011. *(A)* Progression of cell aggregation studies. *(B)* Prototype built from behaviour-based modelling environment.

5 Conclusion

The process described here exemplifies an approach studies, generates, develops and articulates the morphological behaviour of material systems which are defined by the application of tension force. The primary pursuit has been the generation of complex force-defined geometries and calibration of such to physical form. A key to comprehending these procedures is that *form* is not the description of a static object; rather it is the description of the morphology of materiality, geometry and dynamic structural forces. Therefore, it is necessary when investigating form that all of its active agents are inclusive to the generative design process. Further research investigates the computation of force driven adaptations of local topologies, as they form highly differentiated tensioned mesh and surface systems, evolving additional performances in the description of space and environment.

It is important to note the overall placement of this approach within the context of the design and engineering of tension-active structures. As it has been stated, a spring simulation will not mimic actual material behaviour, such as with FEA processes. Additional precision through more accurate levels of computation is necessary. But in accomplishing complex forms, the spring based process provides expediency, relative accuracy, and information beyond geometry to feed expeditiously into further levels of investigation. Research is being conducted to determine and calibrate this alignment between design and engineering simulations.

References

Ahlquist, S., Fleischmann, M.: Cylindrical Mesh Morphologies: Study of computational meshes based on parameters of force, material, and space for the design of tension-active structures. In: 27th eCAADe Conference Proceedings of Computation: The New Realm of Architectural Design , Istanbul (2009)

Ahlquist, S., Menges, A.: Realizing Formal and Functional Complexity for Structurally Dynamic Systems in Rapid Computational Means: Computational Methodology based on Particle Systems for Complex Tension-Active Form Generation. In: Ceccato, C., et al. (eds.) Proceedings of Advances in Architectural Geometry Conference 2010. Springer, Berlin (2010)

Fleischmann, M., Ahlquist, S., Menges, A.: Articulating Boundaries of Computational Meshes. In: Proceedings of the Design Modelling Symposium, Berlin (2009)

Goel, A., Rugaber, S., Vattam, S.: Structure, Behavior and Function of Complex Systems: The SBF Modeling Language. Georgia Institute of Technology (2009)

Gordon, J.E.: The New Science of Strong Materials or Why You Don't Fall through the Floor, Revised edn. Princeton University Press, Princeton (2006)

Kilian, A.: Linking hanging chain models and fabrication. In: Proceedings of the 23rd Annual Conference of the Association of Computer Aided Design in Architecture and the 2004 Conference of the AIA Technology in Architectural Proactice Knowledge Community, Cambridge (2004)

Kilian, A., Ochsendorf, J.: Particle-Spring Systems for Structural Form Finding. Journal of the International Association for Shell and Spatial Structures (2005)

Knippers, J., et al.: Plastics and Membranes Construction Manual: Materials and Semi-Finished Products, Form Finding and Construction. Birkhäuser Architektur (2010)

Linhard, J.: Numerisch-mechanische Betrachtung des Entwurfsprozesses von Membrantragwerken. Dissertation, Technischen UniversitätMünchen (2009)

Moncrieff, E.: Systems for Lightweight Structure Design: the State-of-the-Art and Current Developments. In: Onate, E., Kroeplin, B. (eds.) Textile Composites and Inflatable Structures. Springer, Netherlands (2005)

Synthetic Images on Real Surfaces

Marc Alexa

Abstract. We describe several ways to create real surfaces that display one or more prescribed images formed by different effects on the surface, such as diffuse reflection or self-occlusion. Especially the effect of showing different images for different directions of light is surprising and not commonly seen in the real world. We discuss necessary theoretical limitations, measurements, and the different algorithms. The resulting surfaces have applications in entertainment, fabrication, and architecture.

1 Introduction

The earliest images generated by humans have likely been sculptured in a stone surface, rather than created by painting. For example, sculpting chisel marks on surfaces causes shading due to occlusion. A skilled artist can use this shading to induce an apparent desired texture even on surfaces made of a uniform material. A more refined technique exploits that the reflection of a diffuse surfaces varies with the angle between surface normal and direction to the light source. Thus, varying the height results in varying radiosity across the surface. This effect can also be used to convey a particular image – an approach usually called relief.

In our work we explore similar concepts, albeit controlled with a computer. Our goal is to automatically produce a model of a surface made of a uniform material that can reproduce a desired image. Once we generate a computer model of this surface, we can use automated devices such as computer-controlled CNC milling machines to physically fabricate it. Thus, our aim is to take an analog method that has been used by skilled artists for centuries and to convert it to a digital process and, by doing so, also *extend its scope*.

Marc Alexa
Computer Graphics, TU Berlin, Germany

Fig. 1 An input image and two reproductions in wood, one with regular hexagonal pattern and the other one is an optimized pit distribution

2 Self-occlusion

Our process uses cylindrical pits as basic geometric primitives (although this method can be easily extended to accommodate different primitives as well). We make a simple observation: when we increase the depth of a pit, the average apparent albedo of the surface patch containing this pit decreases. This is because a deep pit generally absorbs more light than a shallow one. One of our goals is, clearly, to optimize the apparent match between the computer prediction (or simulation) and the appearance of the real surface. For predicting the dependence between hole depth and albedo we use a data-driven solution (see below). The resulting look-up table allows us to go directly from a desired image to a pitted surface based on a regular grid and thus a desired surface model. However, we find that it is also beneficial to investigate irregular hole patterns. Irregular grids allow us to improve the resolution mismatch between the number of pits in the fabricated surface and the number of pixels in the input image by aligning pits with the dominant image features, see Figure 1.

We believe that our approach has several interesting and unique properties. First, the process is completely ink-free – we can embed arbitrary images into almost any surface and without using any dye or additional material. The approach is relatively inexpensive (for example in injection molding), simple to implement, and works well on large-scale surfaces. The produced surfaces are durable and integrate well with current materials. We also believe that these surfaces have a very unique, aesthetic feel and they can be appreciated by many observers. Finally, this method could also be used in conjunction with surfaces with non-uniform albedo (e.g., to increase the dynamic range of the medium).

2.1 Data-Driven Pit Model

We would like to produce the best possible match between the desired image and the radiance values observed by the viewer. While these values depend on the viewer's

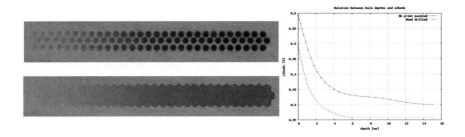

Fig. 2 A drilled calibration pattern in a wood sample (above) and the fitted hexagonal patches (below) on the left. The resulting curves for this sample as well as a 3D printing polymer are shown on the right.

position and the illumination environment we have found that the exact position of the viewer is not critical in practice.

First, we manufacture a calibration pattern for each material type. We use this pattern to determine the dependence of the apparent albedo of a surface patch on the pit depth. Our pattern consists of several rows of equal lines of pits in ascending depths from zero to a maximum depth. We use a white diffuse tent to approximate the uniform illumination. The camera is placed directly above the surface. We also use a material sample with known reference albedo and we estimate the albedo of our surface by comparing it to the reference material.

Next, we process the calibration image in order to compute the mapping between the apparent albedo and the pit depth. In particular, we use a hexagonal box filter that covers a Voronoi region of each pit. Integrating over this region yields the average value or the effective albedo of the pit and its surrounding area. We obtain the absolute value by comparing this value to the calibration white albedo target captured in the same image. We show a sample calibration target and the corresponding processed image with applied filter in Figure 2. Once the average apparent albedos for each pit depth are measured it is straightforward to convert this function to an inverse mapping. As we might not have the measurements for all necessary pit depths, we approximate the intermediate values with linear interpolation (see Figure 2).

2.2 Pit Placement

Given a surface area, the desired pit radius r, and a minimal distance between holes ($> 2r$), the largest number of pits can be placed using a hexagonal grid for the pit centers. However, it is not clear that the largest number of pits necessarily yields the best reproduction of a given image function $I(x,y)$. And, indeed, we find that the uniform pit distribution of the output image often exposes clearly visible and distracting aliasing artifacts along image edges.

We wish to quantify the difference between the image conveyed by the pitted surface and the image I. For this, we represent the input image $I(x,y)$ with a set of discretely placed samples S. The set S contains elements $s = \{s_c, s_v\}$, where s_c is an image sample coordinate and s_v is the corresponding image value. In practice, we use a grid that is a few times *finer* than the input raster image, i.e. we represent each pixel in the input image by a set of 2 by 2 to 5 by 5 samples.

Now we can associate each pit p to the discrete representation of the Voronoi cell S in the image, i.e. S_p contains the samples s for which s_c is closest to p_c. Based on this association we define the average error as

$$E = \sum_{p \in P} \frac{1}{|S_p|} \sum_{s \in S_p} \|p_v - s_v\| = \frac{1}{|S|} \sum_{p \in P} \sum_{s \in S_p} \|p_v - s_v\| \tag{1}$$

In order to minimize this error we optimize the pit centers with a weighted variant of Lloyd's method [5]. This algorithm iterates between the definition of the cell associated to each pit and then updating the pit position to the centroid of the cell. The basic observation is that we need to bring each pit into a position so that its value p_v is close to all associated values s_v, i.e. ideally each pit represents a homogeneous area of the image. This idea is reflected in the definition of the two steps below, taking into consideration not only the positions but also the values of the pits and samples.

In the first step, we use the following cost function $c(s,p)$ to associate the samples to a pit position:

$$c(s,p) = \|s_c - p_c\| + \sigma_v \|s_v - p_v\|. \tag{2}$$

The weight σ_v corresponds to the importance of image values compared to the Euclidean distance. Associating all samples to all pits requires $O(|S||P|)$ for a simple scan over all points. We use Dijkstra's algorithm to speed up this process.

In the second step of the algorithm, each pit center is moved towards the weighted centroid C_p of its associated sample points computed in the first step. Given the centroid and the corresponding pit p we update each pit position and its value as follows:

$$p_c \leftarrow p_c + \delta(C_p - p_c), \tag{3}$$

where parameter δ controls the step size towards the weighted centroid. We set δ to less than 0.1 in order to improve convergence. Furthermore, the pit center update step might place the centers too close to each other for a given pit radius and minimal wall size. We simply disallow these types of updates.

2.3 Results

The final self-occlusion surface is described as a set of pit centers. We have compared the regular (hexagonal) pit distribution to the irregular distribution for several examples using the error functional in eq. 1. As intended, the irregular placement minimizes the error along the contrast edges in the input image. We show a comparison of regular vs. irregular distribution for font reproduction in Figure 3.

Synthetic Images on Real Surfaces

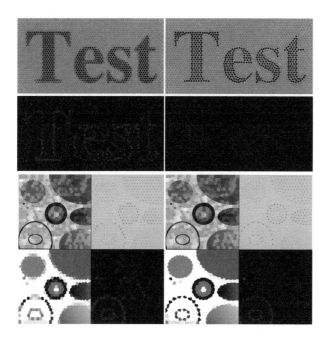

Fig. 3 Comparison of the uniform hexagonal tessellation and the optimized irregular placement. The right images show how the optimization advances from an initial state.Clockwise starting from images in upper left in each set: regions corresponding to each pit, centroid distribution, average error for each region, average region values.

Figure 3 depicts the results of the algorithm for computing the irregular pit distribution. The left image set shows the initial state where the pit centers are still placed as the regular grid. Due to the regular pit placement the positions of the pits are not aligned with the image features leading to higher errors. On the right side we show the state after 400 iterations. In this case the sites are now placed aligned with the image features leading to smaller errors on many edges and visually smooth shapes.

We have experimented with fabricating the surfaces using two different methods: by drilling holes using a CNC machine and by using a 3D printer. One could also manufacture surfaces and objects using injection molding methods. These methods could deliver at least an order of magnitude higher resolution compared to the methods we have used. The only downside is the high cost of manufacturing the molds, which is typically amortized by mass-producing many copies.

3 Diffuse Reflection

Reliefs are usually created by compressing the depth image of real 3D geometry. This technique has been used for centuries by artists, and has recently been

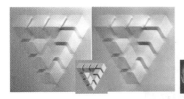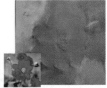

Fig. 4 Given arbitrary input images (inset), we compute reliefs that generate the desired images under known directional illumination. The inputs need not depict real surfaces (left), change in light direction can be exploited to depict different images with a single relief surface (middle), and if the light sources are colored, even a white surface can generate a given color image (right).

reproduced digitally (see, for example, [6]). These approaches effectively exploit the so-called bas-relief ambiguity [2] in order to predict the image generated by the relief surface under constant uniform illumination and mostly diffuse reflection.

However, the range of images that could be conveyed by a surface is richer than surfaces obtained with bas-relief ambiguity. In this work, we define the following more general problem: given an input image (or a set of images) and the corresponding light directions, we would like to obtain a relief such that its diffuse shading approximates the input image when illuminated by this directional lighting.

It turns out that this problem, in general, cannot be solved. Quite simply, there are images that cannot be the result of diffusely shading a surface with constant albedo [4]. The reason for this is that the normals of a continuous surface are not independent – the vector field of normals is necessarily *integrable* or *irrotational*. Without this limitation it would be quite simple to solve the problem of generating a desired image as the reflection of a diffuse surface (disregarding interreflection and self-occlusion): if we fix viewing and lighting directions then each pixel value in the image directly defines the angle between surface normals and light direction.

The main idea and contribution of this work lies in a discrete surface model that effectively alleviates the problem of constrained normals (see section 3.1): each pixel is associated not only with one but with several elements of the surface. Together, the surface elements provide enough degrees of freedom and the different resulting radiance values are effectively integrated for realistic viewing distance.

We show several results (see 3.3) of a optimization procedure based on this model. When the same image is used for two or more light source positions we get relief results that generate stable images for a wide range of light source directions and viewing directions. This is an interesting result, as the images are not necessarily depicting diffuse surfaces under single light source illumination. We also show that it is possible to generate two different images with a single relief surface.

We present a rather simple approach, taking into account only direct shading, but no interreflection, sub-surface scattering, and other realistic shading effects. Consequently, we believe that this is only the beginning of using surfaces as displays. More discussion and conclusions are presented in section 4.

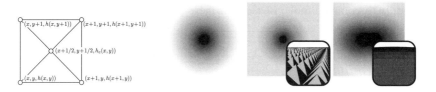

Fig. 5 The discrete representation of a pixel (left) and the creation of reliefs for the canonical "impossible" image from [4]: using pyramids we could create a qualitatively perfect reconstruction of the image (i.e. the gradient images are almost identical), however, they are view sensitive and hard to construct so we have to resort to the smoothed version on the right.

3.1 Discrete Relief Model of a Pixel

We will represent the relief by a discrete surface, fixing the vertex positions in the plane and only adjusting the heights (i.e. a discrete height field). A simple counting argument shows that a regular grid of vertices, associating one quadrilateral element to each pixel, seriously limits the space of images: asymptotically, there is one degree of freedom per element. This is independent of whether we consider non-planar quadrilateral elements or triangulate each element to form a piecewise planar surface. Horn et al. [4] have analyzed the problem theoretically and give canonical images that cannot be the result of shading a smooth surface (see Figure 5 left).

We overcome this limitation by adding an additional degree of freedom: each element of the relief surface corresponding to a pixel in the images is modeled as four triangles forming a square in the parameter domain (see Figure 5). The four triangles will have different normals and, thus, different foreshortening. When viewed from a distance a human observer will perceive the aggregate radiance of all four facets. Note that the one additional vertex in the center of each element adds exactly the one necessary degree of freedom per element – in this model, each element corresponding to a pixel in the image offers (asymptotically) two degrees of freedom.

Assume for the moment that two light source directions are fixed, and two discrete gray level images are given. We further assume the viewer is distant relative to the size of the relief. The goal is to create a discrete relief surface that reproduces each pixel's intensity as the integrated radiance of the corresponding 4 triangles.

We would need to compute the radiance of a surface element as viewed from the viewer's direction. First note that the *projected area* of all four triangles is $1/4$. Their areas are different and can be computed as cross products of interior edges in each pixel. For purposes of iterative optimization (see below) we will assume these areas to be constant, which then leads to an expression for the radiance of each pixel that is linear in the heights of vertices (for details see [1]).

In Figure 5 we illustrates how much flexibility can be gained from using pyramid shaped elements. As the input image we use a canonical impossible image example by Horn et al. [4]. Then we show what was possible if we used arbitrary pyramids for the reconstruction. Unfortunately, the resulting variation in height makes the result unstable under even slight changes in viewing direction, as many of the triangles are at grazing angles with the light sources. In addition, it would be difficult to machine the surface with common tools. Because of these problems we impose additional smoothness constraints.

3.2 Optimization

The overall goal of the optimization process is to compute a surface model whose diffuse shading reproduces the desired image or images for given directional light sources. In this optimization process, we minimize the squared difference of radiance gradients and the corresponding input image gradients. We also add smoothness and damping terms to get constructable reliefs and a wide viewing angle.

The dynamic range of shading produced by relief surfaces is quite limited. Therefore, it is desirable to tone map the input images as a preprocessing step. We use the image gradient compression method similar to approaches by Fattal et al. [3] and Weyrich et al. [6].

Then, rather than optimizing for the absolute gray levels in the images we optimize to match their gradients as close as possible. In particular, we minimize the squared differences between radiance gradients and corresponding compressed image gradients:

It is also important that each pixel is close to flat. This can be achieved by minimizing the second order differences within a pixel. For center vertices, this term is expressed using a sum of their squared Laplacians. Intuitively, minimizing the Laplacian pulls each center vertex towards the mean height of its neighbors. However, moving the center vertex to the mean height is not enough to keep the geometry of a pixel planar. It is also necessary to consider the second order smoothness of corner vertices of each pixel.

Our optimizations relies on the assumption that the overall height of the desired relief should be small. Consequently, we penalize large height values.

The main difficulty in optimizing the energy terms is due to its non-linearity. However, as mentioned before, we damp the change of triangle areas and then iterate over solutions for standard quadratic expression, in which we assume the triangle areas to be constant. Then the gradient of the energy becomes linear and can be solved for efficiently.

3.3 Results and Applications

We use the proposed optimization method to create example surfaces that convey various input images. In all examples we optimize for two or three directional light sources.

When generating a relief that depicts a single grayscale image under different light directions we optimize using the same input image for all desired light directions. The resulting relief image is usually slightly different for different directional illuminations. However, the image can be clearly perceived even for light directions that have not been part of the optimization. The case of depicting a single grayscale

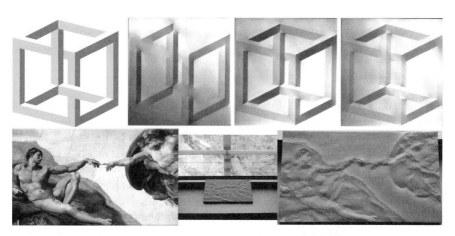

Fig. 6 Top row: the desired image, and the resulting machined relief surface for different lighting directions. Lower row: A relief created from the image on the left. The image is visible despite suboptimal illumination.

image is less constrained and therefore we can afford to make the surfaces relatively flat.

As an interesting example, we have manufactured surfaces that depict "impossible" geometry such as the impossible triangle (see Figure 4) or the famous impossible cube (see Figure 6). For the cube we show several different lighting directions. To show how insensitive the reliefs are to the actual illumination we took a photograph of the relief in Figure 6 under very general lighting conditions, i.e. daylight that has a dominant direction because it comes through a window.

We have also created several reliefs that depict two different images when illuminated from two distinct light directions. We observe that high contrast edges that are present in the input images are visible in the relief surface regardless of the light direction. While unfortunate, this cannot be avoided when the output surface is reasonably smooth. This is because a high contrast edge in the image corresponds to a line of high curvature in the fabricated surface. This edge is going to be visible unless the direction of this curvature maximum is aligned with the projection of the light direction. This requirement is violated by any edge that is curved almost everywhere. Therefore, input image pairs that can be reliably reproduced have contrast edges in similar places or in regions where the other image is stochastic. For example, in Figure 4 we show a result for which we warp two different face images such that the eyes, nose, and mouths coincide.

Since our method allows depicting two or more similar input images we attempt to reproduce colored images as well. Given the colors of the light sources, we project a color image into these primaries yielding a set of input images. These images are used as input to our optimization process to obtain a surface which approximates the color image when illuminated by this colored lighting. We believe that these reliefs are the first to reproduce color images on white diffuse surfaces. The example in Figure 4 uses two primaries, a red and a green light.

4 Conclusion

This paper discusses solutions for embedding arbitrary images in a material with a uniform albedo. We exploit shading either due to self-occlusion or due to a diffuse reflectance function to design, simulate, and fabricate surfaces that induce desired images.

We believe that there are many possible avenues for future work. In the case of pits, we would like to explore pits with different diameter (as opposed to fixed diameter pits) and pits with more general cross-sections. This would allow us to increase the dynamic range (as the wall area is fixed, the ratio of the pit area to the total area is increasing with larger diameter). Pits with different cross-sections could improve image reproduction quality.

For the reliefs, which are based on reflection, one may use arbitrary BRDFs, consider subsurface scattering, or interreflections during the optimization. It is would also be possible to optimize reliefs for point light sources instead of directional lights. This would extend the applicability of our method as many real world lights can be approximated using a point light source model. Furthermore, using point light sources might also add an additional degree of freedom and allow generating more than two images with a single relief surface.

In general, given that the many surfaces are fabricated with computerized tools, it is conceivable that their albedo can be controlled as well. This would make the optimization process less constrained and it would very likely lead to better visual results. Another direction is to modify arbitrary 3D models (instead of planar surfaces) in order to enrich them with arbitrary grayscale textures.

References

1. Alexa, M., Matusik, W.: Reliefs as images. ACM Trans. Graph 29, 60:1–60:7 (2010), http://doi.acm.org/10.1145/1778765.1778797
2. Belhumeur, P.N., Kriegman, D.J., Yuille, A.L.: The bas-relief ambiguity. Int. J. Comput. Vision 35, 33–44 (1999), http://portal.acm.org/citation.cfm?id=335600.335604, doi:10.1023/A:1008154927611
3. Fattal, R., Lischinski, D., Werman, M.: Gradient domain high dynamic range compression. ACM Trans. Graph 21(3), 249–256 (2002), http://doi.acm.org/10.1145/566654.566573
4. Horn, B.K.P., Szeliski, R.S., Yuille, A.L.: Impossible shaded images. IEEE Transactions on Pattern Analysis and Machine Intelligence 15(2), 166–170 (1993)
5. Lloyd, S.P.: Least squares quantization in pcm. IEEE Transactions on Information Theory 28, 129–137 (1982)
6. Weyrich, T., Deng, J., Barnes, C., Rusinkiewicz, S., Finkelstein, A.: Digital bas-relief from 3d scenes. ACM Trans. Graph 26(3), 32 (2007), http://doi.acm.org/10.1145/1276377.1276417

Modelling Hyperboloid Sound Scattering
The Challenge of Simulating, Fabricating and Measuring

Jane Burry, Daniel Davis, Brady Peters, Phil Ayres, John Klein, Alexander Pena de Leon, and Mark Burry

Abstract. The *Responsive Acoustic Surfaces* workshop project described here sought new understandings about the interaction between geometry and sound in the arena of sound scattering. This paper reports on the challenges associated with modelling, simulating, fabricating and measuring this phenomenon using both physical and digital models at three distinct scales. The results suggest hyperboloid geometry, while difficult to fabricate, facilitates sound scattering.

1 Introduction: The Consecration of the Sagrada Família

The construction of the interior of Gaudí's Sagrada Família church was completed in 2010. One consequence of the sudden transition of this interior after 126 years, from a site of relentless construction activity to a church, is the first full experience of the acoustics of the space, released from its forests of scaffolding, under the influence of choral, instrumental and clerical presence. Anecdotally, the musicians report the church has a very diffuse acoustic without the long reverberating echoes and islands of intensity characteristic of similar vast volumes constructed

---

Jane Burry · Daniel Davis
Royal Melbourne Institute of Technology,

Brady Peters · Phil Ayres
Center for Information Technology and Architecture,

John Klein
Zaha Hadid Architects

Alexander Pena de Leon and Mark Burry
Royal Melbourne Institute of Technology

from reflective surfaces. Just as the effect of the intersecting doubly curved hyperboloid surfaces of the nave walls is to diffuse the light that washes over them, so too it seems the use of the hyperbolic surfaces scatters the sound in an ecclesiastically novel way.

Fig. 1 Sunlight washing over hyperboloids in the Sagrada Família

2 Building the Invisible

In response to the SmartGeometry 2011 design challenge to 'build the invisible' we held a workshop - *Responsive Acoustic Surfaces* - to investigate further this claim that the hyperbolic surfaces of the Sagrada Família were responsible for creating the unique and diffuse soundscape within the church. As the hyperbolic surfaces of the Sagrada Familia are constructed from stone, which is a sound reflective material, sound absorption was not considered as part of the design problem and it was determined that sound scattering was the most relevant parameter to study. However, compared to sound absorption and reflection, sound scattering proved to be a most challenging acoustic attribute to simulate and measure.

We planned to do this through acoustic testing of large scale prototype walls configured as fields of subtracted hyperboloids. The experience of the group, in particular Mark Burry, of modelling Gaudí's architecture for the Sagrada Família church led to a familiarity with the 3D fourth order curves of intersection between neighbouring hyperboloids of revolution of one sheet. The process of fabricating assemblies of these mathematical surfaces, using gypsum plaster, was also known from the approach to the design of the church through the construction of 1:25 and 1:10 plaster models.

The difficulty of understanding the intersections between hyperboloids distributed on a curved wall necessitated, prior to constructing the 1:1 prototype, the production first of digital models and later 1:10 rapid prototyped plaster models. This paper will focus on the design of the various modelling tools, models, simulations and measuring tools developed to produce and test these prototypes, and report on the interim acoustic outcomes.

3 Digital Modelling

The production of the digital model of the hyperboloid wall, encountered three primary challenges:

- It is computationally expensive to calculate the intersections of multiple hyperboloids but it is vital to understand the nature of these intersections.
- The construction logic required planar intersections, which is a constraint not suited to traditional modes of parametric design or optimisation, nor mechanically unassisted manufacturing (traditional construction).
- Prior research suggests that constrained variation produces better sound scattering than homogeny, since it has been seen that the relational depth of surface components and amount of surface detail impact the scattering coefficient. The variation of the depth of surface components introduces different phase characteristics of the reflected sound waves and therefore affects the distribution of the reflected sound pressures. (Peters 2010)

There being no existing software package sufficiently endowed to overcome these challenges, the project combined two existing software with two custom designed tools to model the wall composed of many intersecting hyperboloids. The first custom tool was a C++ scripting environment built on top the open source library *Open Cascade* to assist in globally propagating hyperboloids across a surface (See Figure 2). The second was a series of custom components for *Grasshopper*™ and *Rhino 3D*™ to automate the modelling of hyperboloids and to simulate the acoustics of various designs.

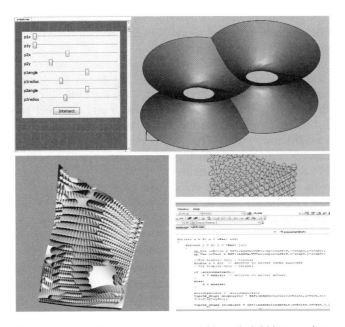

Fig. 2 A sample of the tools used to model hyperboloid intersections

Through trial and error design explorations we discovered a set of geometric axioms that ensured the intersections between adjacent hyperboloids were always planar provided the hyperboloids were identical. These axioms clashed with the requirement for constrained variation and, since they worked additively, it was difficult to develop a scripted or parametric model that also adhered to a predefined shape. For the acoustic tests it was essential the wall conformed to a particular shape and the final design of the wall was developed with scripts, not axioms, although to simplify the construction of the 1:1 prototype all of the hyperboloids were standardised.

4 Physical Modelling and the 1:1 Prototyping

The physical 1:1 construction was conceived of as a single instance or special case of a type of generic output. In early discussion it was posited as a composition derived from discrete sub-units or bricks making up a finite number of geometrical variants in order to be able to model and test a variety of differently composed surfaces. In the weeks leading up to the fabrication and testing - during which the techniques, equipment and knowledge were developed to be able to fabricate the plaster bricks successfully - it became clear that fabricating a family of different 'brick' components would not be feasible within our four day construction window. Instead a single hyperboloid brick was designed such that in multiples they could be stacked together to form a semi cylindrical section. A similarly sized companion wall was constructed from smooth plaster and uses as the control during the sound tests. The overall geometry of both was a semi cylindrical section, chosen because sound originating from the acoustic focus is reflected back to the source, making any change in the reflected sound between the two immediately perceivable by someone standing in the cylinder centre.

Using the same process employed by the Sagrada Família model makers, the hyperboloid of the bricks was shaped as negative plaster moulds by revolving a plasma-cut stainless steel hyperbolic profile about an axis. On top of these plaster master moulds, a hexagonal plywood box was placed, which served a duel purpose as the structure of the brick and as the formwork defining the edge of the brick when the plaster positive was poured. The plaster poured into these hexagonal boxes dried in around 30 minutes, at which point the brick, together with its plywood frame, could be released from the negative.

Since the construction method produced bricks with planar edges, the point where two bricks adjoin - the intersection between hyperboloids - had to be contrived to be planar as well. As was previously discussed, the axioms to ensure the intersections were planer conflicted with the acoustic requirement that the wall be made from a cylindrical section. To meet this requirement, the hyperboloids were propagated across the semi cylindrical wall resulting in an intersection curve that slightly deviated from being planar. This 4mm discrepancy was absorbed by the slight flexibility of the plywood frame when bolted together.

Modelling Hyperboloid Sound Scattering

Fig. 3 Hyperboloid wall (left) and flat wall (right), and corresponding plans

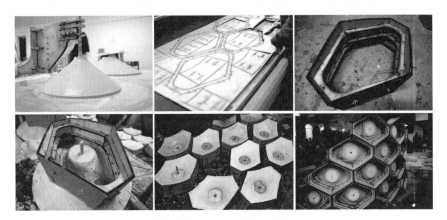

Fig. 4 Top: The hyperboloid mould, laser cutting of hexagonal frame, the frame. Bottom: Hexagonal frame on mould, selection of bricks, assembled wall from behind.

5 Sound Behaviour

Sound scattering is typically measured through the scattering coefficient, which is defined as the amount of sound energy scattered away from the specular reflection direction by a rough surface. (Vorlander and Mommertz 2000) This coefficient is frequency dependent and there is a relationship between the geometry and the scattering coefficient at different frequencies. (Peters 2010) During the workshop, participants developed digital models of hyperbolic geometry from which 1:10 scale rapid prototypes were made and then tested in controlled conditions for their scattering characteristics (See Figure 5). (ISO 2004) The use of both scripting and rapid prototyping in developing these models assisted in quickly iterating designs in response to immediate results all within the four day duration of the workshop.

The sound scattering coefficients are necessary to make reliable simulations and predictions of acoustic performance (Christensen and Rindel 2005) While the ISO standard suggests using physical measurement to determine scattering coefficients, it has been suggested that finite different techniques could be used to calculate these virtually. (Redondo et al. 2009) A digital tool was developed using a similar Finite Difference Time Domain (FDTD) algorithm, to simulate and visualize the interaction of sound waves with the hyperboloid components (See Figure 6). These visualizations demonstrate that the depth of the hyperboloid assists in spreading out the reflected sound wave, so the peaks and troughs are less pronounced. This is a promising first step in the digital simulation of sound scattering but more work is needed to derive the scattering coefficient and to calculate the impulse response.

There are pronounced acoustic differences between the smooth 1:1 wall and the 1:1 wall made of hyperboloids. Standing in the acoustic focus of the smooth 1:1 wall and speaking generates a discernible echo, which is seemingly amplified because the cylindrical geometry concentrates the reflected sound back to the focus point. In contrast, speaking from the same spot in front of the 1:1 wall made of hyperboloids produces no audible echo, since the incident sound is reflected by the hyperboloid geometry in many directions and only a portion of the sound returns to the focus point. These perceptual experiences conform to the acoustic measurements of the 1:10 prototypes, and the visualisations of sound scattering using finite difference techniques.

Fig. 5 Graph of scattering coefficient with respect to frequency, and corresponding 1:10 plaster model

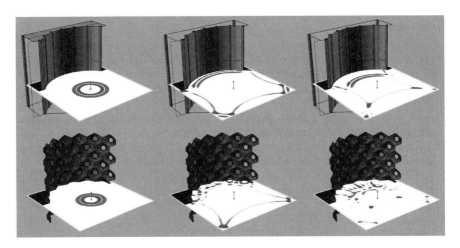

Fig. 6 FDTD Visualization of Scattering from smooth wall (above) compared to hyperboloid wall (below)

6 Conclusions

The *Responsive Acoustic Surfaces* workshop at SmartGeometry 2011 brought together a wide variety of simulation, fabrication and measuring techniques to analyse the relationship between hyperboloids and sound scattering. In producing architecture that performs acoustically, it is necessary that the performance is measureable, and that there is some sort of feedback system that allows designs to be evaluated. In this case the sound scattering performance was evaluated by both physical models and by sound wave visualizations. Both techniques need development, but point to a method to integrate sound performance in parametric architectural tools.

The interim results of the acoustic test indicate the geometry of hyperboloids have a perceptual influence on sound scattering, perhaps accounting for the diffuse acoustic of the Sagrada Família. However the challenges inherit with hyperboloid geometry makes them, at this point in time, impractical sound scattering devices – despite this study's novel innovations in the design, modelling and fabrication of hyperboloids. It is anticipated that future work will identify geometry with similar acoustic properties to hyperboloids, which are more amenable to modern construction techniques.

Acknowledgements. The authors are grateful for the support of: SmartGeometry and the Centre for Information Technology and Architecture (CITA) in facilitating this research; Tobias Olesen, Peter Holmes and Annica Ekdahl for their assistance in running the workshop; and the workshop participants: Adam Laskowitz, Ben Coorey, Eric Turkiemicz, Giovanni Betti, Kathy Yuen, Ralf Lindemann, Robin Bentley and Thomas Hay.

References

Brady, P., Tobias, O.: Integrating Sound Scattering Measurements in the Design of Complex Architectural Surfaces: Informing a parametric design strategy with acoustic measurements from rapid prototype scale models. In: 28th eCAADe Conference Proceedings, Zurich (2010)

Christensen, C.L., Rindel, J.H.: A new scattering method that combines roughness and diffraction effects. Forum Acusticum, Budapest (2005)

ISO, Acoustics - Sound-scattering properties of surfaces -. Part 1: Measurement of the random-incidence scattering coefficient in a reverberation room, ISO, Geneva (2004)

Redondo, J., Picó, R., Avis, M., Cox, T.: Prediction of the Random-Incidence Scattering Coefficient Using a FDTD Scheme. Acta Acustica United with Acustica 95, 1040–1047 (2009)

Vorlander, M., Mommertz, E.: Definition and measurement of random-incidence scattering coefficients. Applied Acoustics 60, 187–199 (2000)

Integration of FEM, NURBS and Genetic Algorithms in Free-Form Grid Shell Design

Milos Dimcic and Jan Knippers

Abstract. Popularity of free-form grid shells grows every day since they represent a universal structural solution for free-form shaped architecture, enabling the conflation of structure and facade into one element [1]. The infinite number of possibilities of generating a grid structure over some surface calls for an automated method of design and optimization, in contrast to the standard trial-and-error routine. This paper presents some results of the comprehensive research dealing with the optimization of grid shells over some predefined free-form shape. By combining static analysis and design software on a basic C++ level we try to statically optimize a grid shell generated over a given surface. Using Genetic Algorithms for the optimization we are able to significantly reduce stress and displacement in a structure, thus save material and enhance stability. The presented method of structural optimization is constructed as a C++ based plug-in for Rhinoceros 3D, one of the main NURBS (Non Uniform Rational B-Splines) geometry based modeling tools used by architects for free-form design today. The plug-in communicates iteratively with Oasys GSA, a commercial FEM software.

1 Genetic Algorithms

Observing examples in Nature, it is obvious that structures with a non-uniform density distribution are ubiquitous. These structures were optimized throughout billions of years of evolution, attempting to minimize material and minimize potential energy, *"for it will profit the individual not to have its nutriment wasted on building up a useless structure"*, as Charles Darwin noticed [2].

The goal of this research was to make a universal method of grid shell optimization, adaptable, easily expandable and with a large number of variables, i.e., with an easy definition of the boundaries and settings within which we want our solution to be generated. Therefore a plug-in was developed so that the user can: **1.** Choose the surface over which the grid will be generated, **2.** Chose the basic pattern of the grid (e.g. Delaunay triangulation [3], quadrangular, Voronoi [3], *Voronax*, etc.), **3** .Set a support combination (e.g. all four edges, two edges, fully restrained, movable, etc.) **4.** Set a load combination (any load combination definable

---

Milos Dimcic · Jan Knippers
Institute of Building Structures and Structural Design,
Stuttgart University, Germany

in FEM software), **5.** Set material properties, **6.** Set cross-section of the structural members, **7.** Define the fitness function (e.g. minimize Von Mises stress, minimize displacement, maximize Load buckling factor, etc.), **8.** Define one or more penalty functions (e.g. limit the length of a member, limit the size of a polygon, limit the stress generated in one member, etc.), **9.** Set GAs parameters (e.g. crossover and mutation probability, number of individuals, number of generations, etc.). Each one of these settings (Fig. 1) can be easily expanded and redefined. When they are chosen the optimization process begins and the algorithm converges toward the best solution for that combination of input settings.

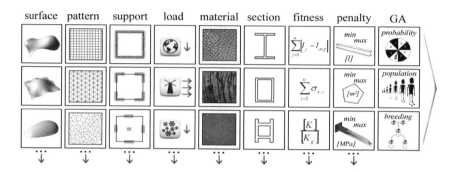

Fig. 1 Input parameters, expandable and changeable

Genetic Algorithms are chosen as a suitable method for multi-objective and highly non-linear optimization. It is a well-known stochastic method and more on the basics of the Genetic Algorithms application can be found in [4]. In the optimization algorithm every individual solution goes through a process of generation and evaluation so that it can be compared to the other solutions and eventually contribute to the finding of the *best fitness* individual. In our implementation, the actions performed on each generated solution are depicted in Fig. 2.

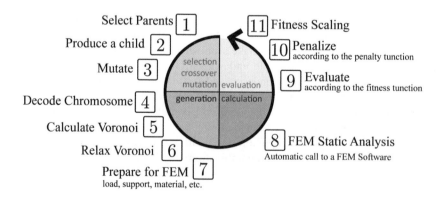

Fig. 2 Basic algorithmic loop for the generation of one grid shell solution

In a standard optimization there are 50 grid shells in a generation, and the process lasts for 400-700 generations, thus sometimes producing more than 30.000 solutions. All the solutions are kept in the specific files that enable their recreation, i.e., extraction and drawing of any of the generated grid shells in the process. We will concentrate on the generation steps (4-7) and the calculation of the structure (step 8).

2 Grid Generation

Genetic Algorithms work with a *chromosome* representation. In this research the chromosome is formed as a string of real-valued numbers which are later on transformed into the *uv* coordinates on the surface and finally into a grid shell. Steps 4-7 cover this process. Although our software can deal with different grid patterns (regular or irregular) the most interesting results come from the optimization with Voronoi pattern. We used the algorithms for the generation of a 2D Voronoi diagram to form a grid over a free-form surface (with a direct xy –uv transformation). Then we applied the Force-Density method [5,6] and adapted (expanded) it so it can be used to *relax* any kind of grid (any kind of graph basically), while keeping it on the predefined surface. This resulted in a new type of structure we named Voronax (Voronoi + Relax), which is created by relaxing a Voronoi grid over a given free-form surface (Fig. 3). Voronax is a foam-like structure, since the principle of its generation is a dynamic search of equilibrium, similar to the process happening within actual foam. Its polygons have much more similar angles and lengths than the Voronoi grid, which is preferable for the grid shell design.

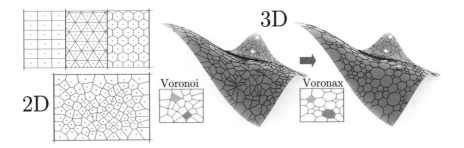

Fig. 3 Left: Voronoi in 2D. Middle: Voronoi over a surface. Right: Voronax

Letting the Voronoi points be the main variables of the optimization process we can apply Genetic Algorithms in order to find their best distribution, i.e., the best disposition of points used to generate the most efficient grid structure according to the defined objective. The user can chose one or more objective functions, i.e., *fitness functions*, and the process will converge toward the optimal solution based on the single- or multi-objective optimization algorithms.

3 Evaluation

Steps 8-11 in Fig.2 refer to the automated FEM analysis of the generated solution. Namely, each generated grid shell is prepared for the static analysis automatically, according to the user defined parameters (e.g. load, support, cross-section, material, etc.). The FEM software is called, structure is calculated and the data is extracted (e.g. stress, displacements, load buckling factor, etc.) according to the defined objective function. In this way a very fast evaluation (less than 1 second per individual solution) can be accomplished and tens of thousands of solutions can be generated and evaluated within several hours.

4 Optimization

Two examples will be used to demonstrate the effectiveness of the optimization process. At the beginning of each optimization we select a free-form surface that will be used as the basis for our grid shell. Then, some of the basic parameters are set: 1. The load combination, 2. Where and how the structure is supported, 3. Cross-section of the structural members (which is the same for all elements, since our goal is to find the best geometrical disposition of members and not do the cross-sectional optimization), 4. Choose the material, 5. Define the fitness function (e.g., minimize stress, minimize displacement, maximize Load buckling factor, etc.), etc.

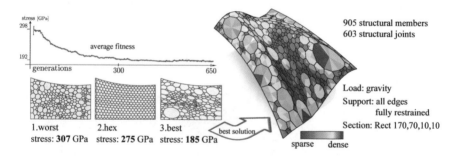

Fig. 4. The Genetic Algorithm based Grid Shell Optimization - Gravity

At the end of the optimization we can analyze the results by looking at the progress graphs and looking at the worst and best generated solution in comparison to the standard, uniform one. The number of structural members and joints in all of the generated solution remains the same. Therefore the basic task can be defined as the improvement of static efficiency while keeping the mass of the structure the same. We try to achieve that by geometrical and topological optimization of the grid, and the color analysis (on the right side of the Figs. 4 and 5) gives us a better insight in what the optimal gird density should look like. The fitness function used in both examples is the minimization of Von Mises stress.

For each structural member in a grid, Von Mises stress is calculated at both of its ends and summed up for the entire structure. It can be seen how the total amount of stress in the worst and best generated solution goes from $307 GPa$ to $185 GPa$ in the first example and from $113 GPa$ to $38 GPa$ in the second example. For comparison purposes, a hexagonal structure is used as the uniform version of the Voronax grid. The reason for this is that Voronax keeps the topology of the Voronoi structure after relaxation, which means that on average its polygons have ~ 6 edges [7] and joints have a 3-member connection (as in a hexagonal grid). The *total displacement* is also used as an indicator of the efficiency of the optimization process. Namely, all displacements from all of the structural joints in a grid shell are summed up and used for comparison. The best generated solution has almost six times smaller amount of total displacement than the worst one (Fig. 5).

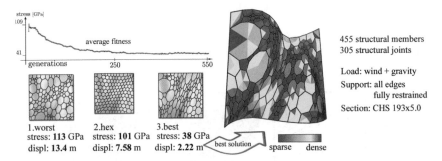

Fig. 5 The Genetic Algorithm based Grid Shell Optimization – Wind +Gravity

5 Interpretation

The results can be taken for granted, but they can be used in order to realize what the optimal grid density looks like so we can design our own statically efficient grid shell. In Fig.6 is an example of how the grid density information can be used to design a quadrangular grid structure. With the similar number of structural members and joints we can lower the stress and displacement by following the *advice* from the Genetic Algorithms.

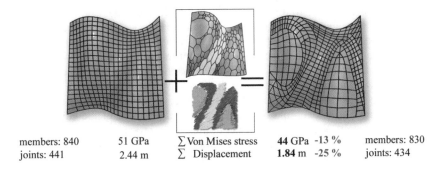

Fig. 6 The density information used for the design of a quadrangular grid

However, when dealing with quadrangular and triangular structures, grid density is only part of the solution. Regular grids are characterized by distinct *structural paths* formed by the connected members. Those paths play as important role as the density in the static efficiency of the grid shell. Therefore, our advice is to combine density information and stress trajectories in order to accomplish the optimal design.

6 Density and Trajectories

By following the maximal Von Mises stress trajectories, and designing a grid so that the structural paths are formed according to those trajectories, we can design a grid shell with better performances. The comparison between a uniform and the optimized quadrangular grid is depicted in Fig. 7, where the optimized solution has 40% smaller amount of total Von Mises stress and 3 times smaller amount of total displacement. Both grids have the same number of joints and members.

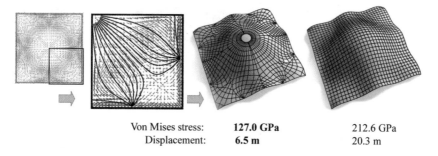

| | Von Mises stress: | **127.0 GPa** | 212.6 GPa |
| | Displacement: | **6.5 m** | 20.3 m |

Fig. 7 The use of stress trajectories for an efficient grid shell design

7 Voronax

Voronax pattern is a very natural structure. It uses the principle of relaxation to find an equilibrium thus providing grid structures very similar to the ones in Nature. By combining it with the Evolution principle we can analyze whole generations of solutions in order to understand better the effects of the optimization and the directions that should be followed. For example, if we take one of the advanced generations and put all of the generated solutions one behind another, we can get a nice picture of the intention of the optimization process. To illustrate this, in Fig. 8 there are two vertical walls used for the grid shell generation. They differ in the way the grid generated over them is supported. Namely, the red areas represent parts of the surface edges where the generated joints are fully restrained – in the corners on the example on the left and in the middle of the edges in the example on the right. We can see how the optimization process shows different intentions in these two cases and how easy interpretable those intentions are. It can be seen how, in order to design a grid shell with lower stress, we should follow the O-shaped pattern of the enhanced density in the case on the left, and an X-shaped pattern in the example on the right.

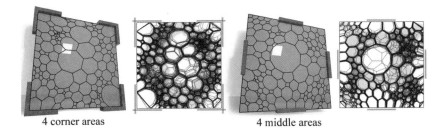

4 corner areas   4 middle areas

Fig. 8. The *intention* of the optimization process with different support combination

8 Future

The final step that has to be performed in order to have a complete tool for a statically efficient grid shell design is the automated geometry generator that would create an optically and statically acceptable grid according to the information provided by Genetic Algorithms. (For example, when we want to use the Voronax optimization to create grids with different patterns). So far we used that information and tried to design the structure *manually*. This is a very big step because without the optimization we wouldn't know how to design it. A limited trial-and-error approach is not enough for the complexity we are confronted with nowadays. In the future the process of grid generation will be automated thus bringing the free form grid shell design to a new level.

References

[1] Kolarevic, B.: Architecture in Digital Age – Design and Manufacturing. Taylor & Francis, Abington (2005)
[2] Darwin, C.: On the Origin of Species By Means of Natural Selection. Kindle edn. (1998)
[3] De Berg, M., Cheong, O., Van Kreveld, M., Overmars, M.: Computational Geometry. Springer, Berlin (1997)
[4] Goldberg, D.E.: Genetic Algorithms in Search. Optimization and Machine Learning. Addison-Wesley, Reading (1989)
[5] Gründig, L., Moncrieff, E., Singer, P., Ströbel.: A History of the Principal Developments and Applications of the Force Density Method, Method in Germany 1970-1999, In: IASS-IACM, Chania-Crete (2000)
[6] Linkwitz, K., Schek, H.-J.: Einige Bemerkungen zur Berechnung vonvorgespannten Seilnetzkonstruktionen. Ingenieur-Archiv. 40, 145–158 (1971)
[7] Sack, J.R., Urrutia, J.: Handbook of Computational Geometry. North-Holland, Amsterdam (1999)

SOFT.SPACES
_Analog and Digital Approaches to Membrane Architecture on the Example of Corner Solutions

Günther H. Filz

Abstract. The desire for new mostly fluent "soft spaces" in architecture cannot be overseen any more. Therefore new tools and approaches are tried out to create architecture with special spatial qualities. In combination with common building-technologies and considered as architectural elements, spatially curved membrane structures and especially anticlastic Minimal Surfaces offer one possible approach to this topic[1]. This paper presents the result of the research on Minimal Surfaces of membrane corner solutions representing one part of a comprehensive basic research in this field. In connection with the exploration of corner solutions, the meaning and potential of analog and digital approaches became obvious. Depending on the potential of the selected tool, the desired results or the purpose of the research one of these strategies will be preferred. Further on this paper partly reveals new correlations between Minimal Surface and boundary conditions and so far unknown rules in its selforganizing processes. Case studies document new capabilities in designing and creating space in architecture. Latest approaches are dealing with alternative boundary conditions, case studies and with software implementation in terms of scripting found rules[2].

1 Introduction

Based on the research of Frei Otto and his team at IL (University of Stuttgart) and the resulting exceptional pioneer constructions, building with textiles as an alternative to traditional materials like wood, stone, steel, glass, and concrete was rediscovered during the last decades. For their structural advantages, prestressed, spatially curved Membrane Structures were up to today mainly used for wide span, lightweight-structures. The formal/architectural point of view of this kind of structures was neglected. Representing the contrast to our right-angled, conventionally built environment, the desire for fluent "soft" spaces in architecture cannot be overseen any more. The possibility to create light and fluent spaces as a

---

Günther H. Filz
Institute for Structure and Design, Faculty for Architecture, University of Innsbruck, Austria

symbiosis of form and structure with anticlastic Minimal Surfaces offers new qualities and chances for today's architecture.

2 Subject

The fluent forms of Minimal Surfaces are fascinating by their variety, structural performance, reduction to the minimal in terms of material use and resources and their special fashion-resistant aesthetics. Together these parameters represent the common basis of a potential design or design concept and characterize its grade of sustainability. Seen as an element in the design of architecture these anticlastic, fluent forms caused by structural conditions, follow the rules of FORMFINDING in its initially (by Frei Otto) defined sense. Very often we misuse the term „formfinding". What Architects mostly mean and do is a man controlled process of SHAPING - a process that happens on a consciously controllable and formal level. In contrast to the man-controlled process of shaping, forms that are arising from selforganizing processes can only be influenced by the design of their boundaries. The form itself can only be found and represents the result that cannot be manipulated. Therefore, the architect finds himself in the unusual position of a creative "formfinder" instead of the "shaper". On the other hand there is a very restricted number of digital tools is available only that are able to calculate these complex selforganizing forms and geometries in terms of Minimal Surfaces. In some cases the analog formfinding still can be very efficient, precise and make sense.

3 Objectives

Since selforganizing processes follow precise rules and contain optimization by their nature descriptions and especially in architecture illustrations of these rules can be used as design tools. This paper presents the result of the research on Minimal Surfaces of membrane corner solutions representing one part of a comprehensive basic research that considers membrane forms as elements in architecture and in combination with common building-technologies. To find out about the chances for an architecture between „hard" and „soft" morphology, basic research on the systematic determination of very different boundaries – the interface between membranes and common construction technologies – enables the opportunity to analyze anticlastic Minimal Surfaces regarding form and curvature. Vice versa we get an idea of the correlation between 3d-curvature, deflection and determined boundary and further on an idea of formal and structural behavior. With the investigation on membrane corners, also the meaning of the applied method and the selection of the appropriate examination tool was arising very signifycantly.

4 Special Specifications

4.1 Minimal Surface

All experiments are restricted to forms that are deriving from the results of soap-film models – the Minimal Surface. As long as boundary conditions are not changed, Minimal Surfaces can be arranged as a unity arbitrarily in space without changing its form/geometry.

4.2 Interface

Linear, maximal 2dimmensionally curved, bending resistant, line supported boundaries turned out to be the ideal interface between membranes and common construction technologies.

4.3 Membranes as an Integrative Element

Membranes are seen as an integrative component of architecture and are directly connected to other elements of common construction methods. In terms of structural effectiveness the surfaces themselves are considered to be highly efficient by their spatial curvature but not to be load bearing elements for other structural members.

5 Methods

Today the analog approaches in terms of physical (0) and soapfilm models (0) seem to be replaceable by digital experiments (0, 0). Usually digital models are essential for the analysis and evaluation of forms (section curves, their diagrammatic overview, analysis of angles in space,...) of Minimal Surfaces. In the context with this research the assessment and visualization of the Gaussian curvature, which were adapted especially to this research, made it possible to compare and to draw one`s conclusions on different forms and their structural behavior. In terms of selforganizing forms and architectural relevance digital tools hit their limits when even boundary conditions are subject of these selforganizing processes (see 0.1). Here, analog approaches can be the solution.

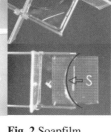

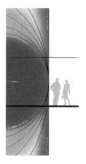

Fig. 1 Physical case study model

Fig. 2 Soapfilm experiment for free corner and selforganizing boundary (S)

Fig. 3 Digital model of horizontal 90° Corner and Visualization of Gaussian Curvature

Fig. 4 Special unification of assessment and visualization of the Gaussian curvature for this research

6 Membrane Corners _ Investigation and Results

Regarding corner solutions with Minimal Surfaces, following boundary conditions are presented with this paper: The right-angled horizontally (0), the right-angled, vertically arranged corner (0) and the free corner (0) which describes a special case because analog tools could deliver satisfying results only.

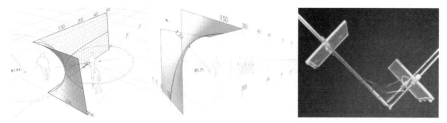

Fig. 5 Horizontal Corner   **Fig. 6** Vertical Corner   **Fig. 7** Free Corner

6.1 Right-Angled, Horizontal Membrane Corner

All executed experiments with horizontal right-angled corners [EH] show almost constant surface curvature (0) and deflection in the area of the corner (0). This happens independently form the leg length and from being arranged symmetrically or asymmetrically. The section lines of digital models are congruent (0, 0). Leg length being shorter than the height cause surfaces with little anticlastic curvature. Surfaces of maximum spatial curvature in all areas can be achieved with a ratio 1/1 to 3/2 of leg length/height. Increased leg length causes areas with little anticlastic curvature at the end of the legs. At the same time the smallest circle of curvature that can be found on the surface is shifted towards the smaller leg (0).

Fig. 8 Horizontal corner with ratio of 1/1/1 (leg/leg/height)

Fig. 9 Horizontal corner with ratio of 1/1/1 Gaussian Curvature

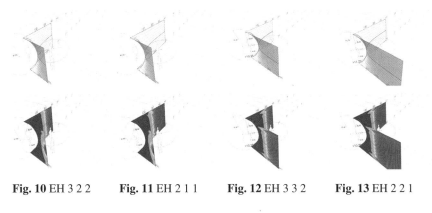

Fig. 10 EH 3 2 2   **Fig. 11** EH 2 1 1   **Fig. 12** EH 3 3 2   **Fig. 13** EH 2 2 1

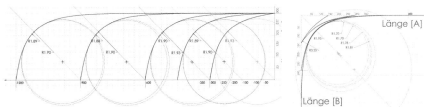

Fig. 14 Horizontal section lines and circles of curvature of right-angled horizontal membrane corner with asymmetrical leg length in comparison

Fig. 15 Horizontal section lines „Horizontal Membrane Corner" – in comparison

6.2 Right-Angled, Vertical Membrane Corner

The configuration of the vertical, right-angled corner [EV] can be used to explore different element length [EL] or different wing length [FL] and their effect on the spatial curvature of the surface. The analysis of section lines, circles of curvature and Gaussian curvature illustrates the interrelationship of surface and boundary proportions. For predominantly curved surfaces these proportions can be located at a ratio of 1/1/1 (element length/height/wing length) whereby even distribution of curvature and harmonic, fluent transitions of surface curvature can be achieved (0,0).

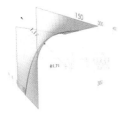 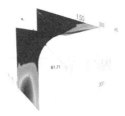

Fig. 16. Right-angled, vertical corner with ratio of 1 1 1 (leg/leg/height)

Fig. 17. Right-angled, vertical corner with ratio of 1/1/1 Gaussian Curvature

6.2.1 Vertical Membrane Corner – Variable Wing Length [FL]

Variable wing length (0, 0) cause change of form of Minimal Surfaces until the wing length is 1,5 times longer than height. From this point the Minimal Surface stays constant in terms of form and curvature. Further increasing of wing length leads to flattened areas at the end of the wing. Wing lengths that are shorter than the height generate strong anticlastic curvature in the area of the corner but the vertical part of the surface looses spatial curvature at the same time.

6.2.2 Vertical Membrane Corner – Variable Element Length [EL]

Elongating the element length (0, 0) means a decrease of curvature in the midspan of the element, while the strong anticlastic curvature in the corner region stays unchanged.

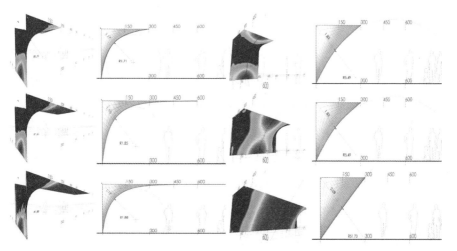

Fig. 18 Vertical corner [FL] variable length, Gaussian Curvature and longitudinal section

Fig. 19. Vertical corner [EL] variable length, Gaussian Curvature and longitudinal section

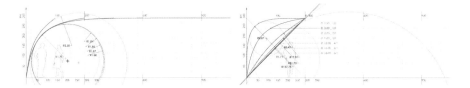

Fig. 20 Overview of sections of vertical corners [FL] with variable wing length

Fig. 21 Overview of sections of vertical corner [EL] with variable element length

6.3 Free Membrane Corner

All boundaries in the course of the investigation on horizontal or vertical corners were supposed to be linear and bending resistant. By the use of flat surfaces instead of linear leg ends freely selforganized boundary curves can be found. Instead of digital models the forms of soapfilm and its free boundary curves were examined for these boundary conditions having different angles at the corner as well as symmetrical or asymmetrical leg lengths (0to 0).

Fig. 22 Free Corner with symmetrical leg lengths and 90°

Fig. 23 Free Corner with asymmetrical leg lengths and 90°

Fig. 24 Free Corner with symmetrical leg lengths and e.g. 135°

Fig. 25 Free Corner with asymmetrical leg lengths and e.g. 135°"

The proportion of 1/1/ (leg length A/ B / Height H) and a right-angled corner causes a displacement (S) of 8.3% at the free boundary curve (0). Longer leg lengths and/or corner angles of more than 90 ° effect smaller displacements (S). Vice versa, the percentage of S exponentially grows by shortening leg lengths or by choosing corners smaller than 90 °. If one leg is longer and the other leg is smaller than the vertical corner distance (H), the displacement (S) in the shorter leg is much small than with symmetrical boundary conditions. This happens even in comparison with symmetrical legs being smaller than the vertical distance (H).

7 Conclusion on Research, Case-Studies and Experimental Structures

The characteristics that Minimal Surfaces can be proportional scaled and that a predefined cutout of Minimal Surface keeps unchanged multiplies the possibilities for the design. Using the found rules, case studies give an idea of the infinite possibilities that are open to create very special „soft spaces", with new architectural qualities like shown in case studies[1] and executed and planned experimental structures (0to 0).

Fig. 26 "Cube of Clouds" experimental structure in model and in scale 1/1; exhibited and published at Premierentage 2005, Best of 2005 and Ziviltechnikertagung 2005

Fig. 27 "Cut.enoid.tower" - experimental structure with a height of about 13meters in scale 1/1. The distorted appearance is generated by the interaction of pin-joint columns, which work on compression and different versions of prestressed catenoids.

Fig. 28 "minimal T"– structure shows the possibility to deflect surfaces that were flat before being assembled by using special geometries in arrangement

8 Perspective

Latest approaches are dealing with alternative boundary-conditions and with software implementation in terms of scripting found rules (0). An investigation on the correlation of self organizing forms, their close relation to nature and their aesthetic values also seems to be interesting questions for the future.

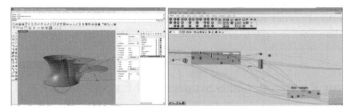

Fig. 29 Grasshopper script "Catenoids between horizontally shifted circular rings"[2]

References

1. Filz, G.: DAS WEICHE HAUS soft.spaces., Dissertation, Leopold-Franzens-Universität Innsbruck, Fakultät für Architektur (July 2010)
2. Filz, G., Maleczek, R.: From Basic Research to Scripting in Architecture. In: Keynote Lecture and Workshop at 2nd International Conference on Architecture and Structure and 3rd National Conference on Spatial Structures, Tehran, Iran, June 15-18 (2011)
3. Otto, F.: IL 18, Seifenblasen. Karl Krämer Verlag, Stuttgart (1988)
4. Mitteilungen des SFB230, Heft7, Sonderforschungsbereich 230, Natürliche Konstruktionen – Leichtbau in Architektur und Natur, Universität Stuttgart, Universität Tübingen, Sprint Druck GmbH, Stuttgart (1992)

Performance Based Interactive Analysis

Odysseas Georgiou, Paul Richens, and Paul Shepherd

Abstract. This paper re-approaches structural engineering through an interactive perspective by introducing a series of tools that combine parametric design with structural analysis, thus achieving a synergy between the architectural shape and its structural performance. Furthermore, this paper demonstrates how the design can be realised into an efficient structural form by applying novel techniques of form-finding through the exploitation of the generated analytical output. The combination of these tools and their parametric control contributes to a new design approach that outrides the generation of single solutions and enables a deeper exploration of the design parameters leading to multiple performance-based outcomes. This paper describes the integration between a Parametric Design software, *McNeals Grasshopper 3D* and a Finite Element Analysis software, *Autodesks Robot Structural Analysis*. The generated synergy between form and structure is demonstrated through a series of case studies through which the interactive control of the parameters the enables the designer to iterate between a range of form-found solutions.

1 Introduction

The rapid advance in CAD technology has enabled architects to overcome the traditional design boundaries and to transform any imagined shape into a persuasive building. In this context, structural design is lagging behind and engineerings engagement with architecture is still restricted. This traditional approach cannot keep up with the modern design process and the engineer is unable to give feedback to the architects design, often stalling the design process. While a large variety of tools serving architectural geometry, such as parametric modelling, is available for use by architects, allowing limitless capabilities and speed in design, the engineering industry remains adherent to traditional structural analysis and design techniques.

Odysseas Georgiou · Paul Richens · Paul Shepherd
Department of Architecture & Civil Engineering
University of Bath, United Kingdom

This paper introduces a novel design procedure through a series of tools that interactively manage and form-find structures. An overview of combining a parametric design software, Grasshopper 3D [1] with a structural analysis software, Autodesk Robot Structural Analysis [2], through the use of computer programming is presented. This combination enables the engineer to retain better control over his designs by employing a performance based approach and it speeds up the design process while allowing for the exploration of new optimum structural solutions [3, 4]. By extending the capabilities of parametric design to include and implement structural analysis, the engineer can move away from the traditional ways of structural thinking and relax the technical boundaries. The results deriving from a structural analysis need no longer be single solutions to problems but parameters that feed into the architectural form and conclude to an optimum shape.

To demonstrate the capabilities of a structural extension to parametric design, two examples of distinct design cases are presented in this paper. Both cases employ the performance based approach by first interactively visualising structural analysis results and then utilising them to iterate between a range of structurally optimum solutions which respond to the designers control.

2 Interactive Structural Modelling and Analysis

The interactive framework that enables a performance based analysis was achieved by linking together McNeals *Grasshopper 3D* and Autodesks *Robot Structural Analysis* using C# programming language. Grasshopper 3D is a plugin built in .NET framework to access McNeals core software, Rhinoceros 3D in order to control and manipulate geometry in a generative manner. The functionality of Grasshopper 3D (GH) can be extended by writing code in C# or VB DotNet programming language to create custom components. In parallel, Robot Structural Analysis (RSA) allows the interaction with other software and the use of its Calculation Engine through an Application Programming Interface (API) [5].

2.1 Methodology

A major part of the process of generating the framework is translating the geometric model to an FEM model. A distinction should be first made between the way that CAD software and FEM software understand and control geometry: when modelling a structure, a set of notions should be taken into account in order to facilitate the process of defining and analysing its performance. These notions can be defined as the structural modelling entities and include nodes, bars and panels and match real building elements as foundations, beams, slabs and so on. Some of these entities also exist in the pure geometric model, sometimes with different naming, and in fact determine a similar notion. For example, two points can describe a line and in the same way, two connected nodes describe a structural bar. What makes a significant difference in the representation of a structural model is the need to attach structural attributes to each of its elements, while the geometric model can be purely described by its topology. In addition, the numbering of each element and its global orientation in relation to its local axis definition are crucial points in the definition of a structural model (Fig. 1). A simple bar element for example, is labelled with a number and is defined by its two interconnected nodes *i* and *j*, each one numbered individually in the global system. The accurate definition of nodes is important in structural analysis since that is where all the calculations occur.

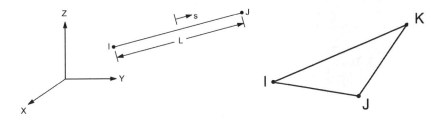

Fig. 1 3D Line element and 3-Node triangular element

Thus, each node that defines geometry in GH needs to be modified accordingly to be read by RSA. For example in order to translate a line to a structural bar, the lines *start* and *end* point coordinates are extracted from GH and are used to define two new nodes for RSA. Those nodes are then used to define a structural bar since there does not exist an implicit way of translating a geometric line to a bar. Each structural member in RSA is defined or controlled using an appropriate interface, which is included in the API library, called Robot Object Model. An interface is a software structure comprising a set of data, defined as attributes or members, and operations that can be performed which are called functions. These interfaces and their functions match the operations that a user follows to model and analyse a structure in the actual software environment. A node for example can be represented by the *RobotNode* interface, which among others includes three real numbers for x, y and z global coordinates. Each node can be managed using the *RobotNodeServer* interface, which for example includes the *Create* function for creating a new node.

3 Applications

A series of application examples are presented in order to illustrate the capabilities of the generated framework.

3.1 Three-Dimensional Grillage Analysis

In this example a three-dimensional structural grillage is linked to a free-form surface that is modelled parametrically in Grasshopper 3D (Fig. 2). By this means, any complex architectural skin can acquire a structure where its performance attributes are visualised interactively responding to the change of geometry.

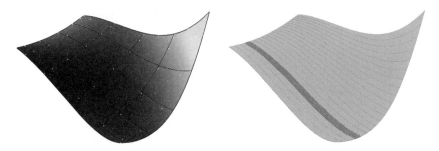

Fig. 2 Free-form surface   **Fig. 3** Surface division in regions

In this case the parametric surface is split into regions according to the users selection and is then transformed to structural data to be read by the analysis software (Fig. 3). These regions also form the loading strips of the structure, according to the load-cases set by the user (Fig. 4).

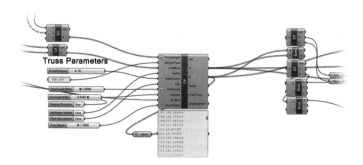

Fig. 4 User parameters in Grasshopper 3D

The edges of the generated regions are subdivided into the desired grillage spacing bays generating structural nodes. The nodes are then offset normal to the surface according to a controllable parameter which defines the height of the grillage. The set of nodes are then connected with structural bars in space, to generate the 3-dimensional grillage. The surface loading is translated to nodal loading to be applied to the structural nodes. The constrain conditions, section types and material properties are also parameters that can be defined in a generative manner by the user, either using numerical sliders or text components. It is the users choice to compute the analytical model inside the Structural Analysis software, in order to perform more rigorous analytical tasks (as this tools capabilities are limited to conceptual analysis), or to visualise the current results in Grasshopper. The following images illustrate the analytical model in RSA including the point loads assigned on to structural nodes and the constrain conditions (Fig. 5) and the visualisation of the stress distribution along the bars of the grillage in Grasshoppers environment (Fig. 6).

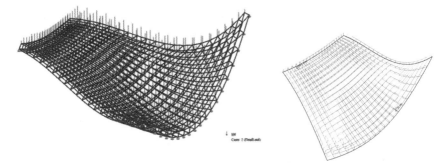

Fig. 5 Analytical model in Robot Structural Analysis

Fig. 6 Visualisation of stress ratios in Grasshopper 3D

3.1.1 Three-Dimensional Grillage Form-Finding

The generated results can further be used to find an optimum form that can coincide with the initial surface. For clarity and in order to prove the feasibility of the proposed optimisation technique, a rectangular planar surface is used instead of the free-form surface shown in the previous paragraph. The stress results are used in an iterative process in order to balance the stress distribution in the bars. Each participating node is translated according to the stress distribution of its interconnected bars following a vector normal to the surface. The user can interactively change the initial parameters to achieve a controlled optimisation result. The optimised shape is then returned into Rhinos graphical environment for assessment or further post processing (Fig. 7 and 8).

3.2 Free-Form Surface Analysis

In a similar manner as in the problem above, a free-form surfaces geometric representation is linked to the structural analysis software through a framework that is able to visually present results as well as to further exploit them for the generation

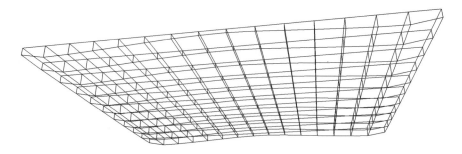

Fig. 7 Optimised grillage geometry in Rhino 3D

Fig. 8 Optimised section through the grillage

of efficient forms. Although the only way to structurally analyse a surface is through the use of discritized or finite elements (planar triangles or quads) it was found practical to first represent it as continuous (i.e. as NURBS) in order to freely control it in three-dimensional space This surface is then translated to an FEM mesh by surface approximation.

The approximate surface or mesh is built in Grasshopper, consisting of a matrix of vertices, edges and faces. This information needs to be decomposed in elements that can be read by the structural analysis component. The mesh vertices are translated in a list of points in 3D space, which are then translated in structural nodes inside the analysis component. The connectivity of each face is then used to create arrays of nodes by selecting the sets that comprise each face. Finite elements can then be created by utilizing the arrays created using appropriate methods in RSAs finite element interface. Structural properties are then to be applied, such as loads, support conditions and material properties. These parameters are also treated in a way that allows them to be controlled parametrically by the user through the GH graphical interface.

After the structural information database is complete, the calculation interface is called, analysing the structure in RSA and its results are sent out to GH for each FEM element. The analysis results are mapped to a faceted surface that was created to resemble the FEM mesh in GH for visualization purposes. Each element values can be tagged accordingly or coloured using a GH *Gradient* component. This generates the interactive representation of the impact of change on the geometry or the constraints of the initial NURBS surface in Rhinos Viewport (Fig. 9).

Fig. 9 Graphical representation of surface stress ratios in Grasshopper 3D

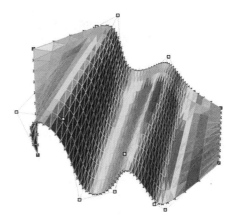

3.2.1 Principal Stress Grid on a Free-Form Surface

An alternative technique for designing freeform grids is introduced in this paragraph. This is based on the directions of principal stresses that occur in a continuum shell. Principal stresses are the components of the stress tensor that occur at each point of a continuum, which are purely axial, consequently their shear component equals to zero. The directions at which these stresses occur are called Principal Stress Vectors. These components share the maximum and minimum stress values and ideally, if a grid is aligned along their directions, it can replace the continuum [6, 7, 8]. (Fig. 10)

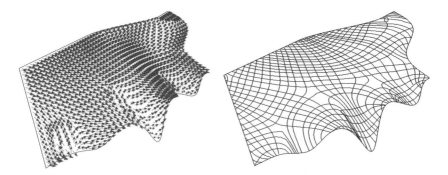

Fig. 10 Principal stress directions   **Fig. 11** Plotted principal trajectories

A series of algorithmic routines are developed inside Grasshopper using C# to plot the principal stress trajectories. These are based on the results obtained using the framework in section 3.2. Some of the basic notions used are described here: The stress data is first mapped on each of the planar surface generated in GH. Selecting an arbitrary face in the mesh, lines are drawn following the direction of the principal stress belonging to the current face. When a line meets a face edge, an intersection occurs which determines the next step and the preferred face at which the plotting would restart. The algorithm goes through all the available mesh faces avoiding the ones already been initialised. The finalised grid is shown in Fig. 11.

4 Conclusions

This paper presented the generation and application of design tools aiming to achieve a synergy between form and structure while at the same time leading to the creation of efficient structures for complex architectural shapes. This was accomplished by employing the power of parametric design combined with structural engineering software, liking them by computer programming. This combination enabled an interactive approach to structural design, a function which currently is sparingly applied for solving engineering problems. The application of the method was demonstrated through two examples: a space grillage and a surface shell. Both cases were interactively analysed responding to the designers controls having their form adapting to multiple variations of external conditions, as supporting conditions loading and material. The presented studies prove possible that optimum structural solutions need not be singular and could allow the designer to iterate between solutions that respond to their performance criteria. Furthermore, this creates the challenge for future adaptation of systems that enable integration of multi-disciplinary design teams to allow interaction of their dedicated parameters for the generation of multi-ojective optimum solutions.

References

1. Grasshopper, Generative modelling for Rhino, http://www.grasshopper3D.com/ (accessed June 25, 2010)
2. Autodesk, Autodesk Robot Structural Analysis Professional, http://usa.autodesk.com/adsk/servlet/pc/ index?siteID=123112&id=11818169 (accessed June 20, 2010)
3. Shea, K., Aish, R., Gourtovaia, M.: integrated performance-driven generative design tools. Automation in Construction 14(2), 253–264 (2005)
4. Shepherd, P., Hudson, R.: Parametric Definition of the Lansdowne Road Stadium. In: International Association of Shell and Spatial Structures, Venice, Italy (2007)
5. Wikipedia, Application Programming interface. In: Wikipedia: the free encyclopedia, http://en.wikipedia.org/wiki/ Application_programming_interface (accessed July 02, 2010)
6. Kotsovos, M.D.: Reinforced Concrete. NTUA publications, Athens (2005)
7. Winslow, P.: Synthesis and Optimisation of Free-Form Grid Structures. Thesis (PhD). University of Cambridge (2009)
8. Michalatos, P., Kaijima, S.: Design in a non homogeneous and anisotropic space. In: Symposium of the International Association of Shell and Spatial Structures, Venice, Italy, December 3-6 (2007)

On the Materiality and Structural Behaviour of Highly-Elastic Gridshell Structures

Elisa Lafuente Hernández, Christoph Gengnagel, Stefan Sechelmann, and Thilo Rörig

Abstract. Gridshell structures made of highly elastic materials provide significant advantages thanks to their cost-effective and rapid erection process, whereby the initially in-plane grid members are progressively bent elastically until the desired structural geometry is achieved. Despite the strong growing interest that architects and engineers have in such structures, the complexity of generating grid configurations that are developable into free-form surfaces and the limitation of suitable materials restrict the execution of elastically bent gridshells.

Over the past ten years, several research studies have focused on methodologies to generate developable grid configurations and to calculate their resulting geometry after the erection process. However, the same curved shell surface can be reproduced by various developable grid configurations which, in combination with their material properties, exhibit different structural behaviours not only during the shaping process but also on the gridshell load-bearing capacity.

In this paper, the structural consequences of the choice of the grid configuration for an anticlastic surface have been analysed by means of FEM-Modelling combined with an geometrical optimisation of the initial bending stresses. In addition, the potential of using natural fibre-reinforced composites as a lightweight and environmentally friendly alternative has been investigated.

1 Introduction

Research works focused on elastically shaped gridshell structures have intensified during the last decade. Since the first large-scale gridshell structure, the Multihalle in Mannheim, Germany (1975), was successfully built, diverse methodologies have been developed which calculate the curved gridshell geometry resulting from the shaping process. The complexity of the calculation relies on the determination of the bending moments induced by the erection process and conditioned by the evolution, deformation and distortion, of the initial grid configuration.

Elisa Lafuente Hernández · Christoph Gengnagel
Department of Architecture, University of the Arts, Berlin, Germany

Stefan Sechelmann ·Thilo Rörig
Department of Mathematics, Technical University of Berlin, Germany

Due to the resultant bending in the profiles, the highest material utilisation in highly-elastic gridshells usually occurs during the erection process. The induced bending moments (1) are directly proportional to the material's Young's modulus and cross-section properties (second moment of area) and inversely proportional to the gridshell curvature (radius of curvature of the grid profiles).

$$M_B = EI / r \qquad (1)$$

In the following figures (Fig. 1), the von Mises stresses of an elastically bent anticlastic gridshell are shown. The gridshell on the left, which is not yet subjected to external loading, already exhibits a material utilisation of about 60%. When an uniformly distributed snow load of 0.9 kN/m^2 is applied to the gridshell, the utilisation percentage increases to approximately 90%.

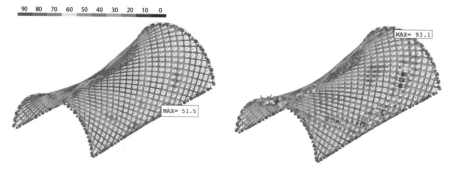

Fig. 1 Von Mises bending stresses of an anticlastic gridshell. Without external loads (left), under uniformly distributed snow (right)

2 Gridshells - Materials of Choice

When choosing materials for highly elastic gridshells, two important aspects should be considered. The first is the material's bending elasticity. The Young's modulus of the chosen material should be sufficiently low in order to minimise bending moments which result from shaping the shell and, consequently, the erection forces during the construction phase as well as the reaction forces at the supports. Conversely, the Young's modulus should be high enough in order to provide the sufficient global buckling stiffness. With high ultimate strains, the profiles are able to be further bent before the maximum stresses are attained.

The second relevant aspect is the material strength capacity. Materials with higher ultimate stresses can resist major bending moments and, hence, are able to adopt a lower radius of curvature and, consequently, are more appropriate for gridshells with strong curvature. Therefore, the choice of the material should not be independent of the gridshell in question, but it should consider its geometric as well as structural requisites.

Moreover, the material selection also affects the definition of the cross-section. While for example timber beams are restricted to solid closed sections, the manufacture of pultruded composite materials offers a great variety of hollow-sections. Here, the shape and wall thickness of the cross-sections can be optimised such that bending stresses during the construction process can be minimised and sufficient global stability of the shaped gridshell can be achieved.

2.1 Existing Materials

The Multihalle in Mannheim, Germany (1975) [1] and the Weald and Downland Museum [2] in Sussex, Great Britain (2002) were both pioneering examples of gridshell structures and both made use of timber, hemlock and oak. With a Young's modulus of about 10 GPa, a bending strength of about 30 MPa and an ultimate strain of about 2%, these timber materials required two profile layers in each grid direction. During the construction process, each layer was bent independently from each other, so that the bending stiffness of only one layer had to be overcome. Once the gridshell was shaped, shear blocks were added between the two layers, in order to transfer shear forces between them. With it, the bending stiffness of both layers together was activated providing the structure with higher rigidity and global stability.

Emerging composite materials such as glass fibre-reinforced plastics (GRP), with higher ultimate strength properties (300 - 400 MPa) and modulus of elasticity (20 - 40 GPa) than timber, are able to afford more rigidity to the structure, so that, instead of two layers of profiles, only one in each grid direction is needed, which facilitates and accelerates construction. The first studies on GRP gridshells were carried out by the Institut Navier, France (2005) [3]. During the shaping process, the bending stresses tend to be higher than by timber, as GRP has a higher rigidity, but this is compensated by a higher ultimate strength, which allows a further bending of the profiles before the maximum stresses are attained. Another advantage of composite materials is that arbitrary profile lengths can be manufactured thanks to pultrusion technology and thereby weaknesses otherwise caused by nodal connections can be avoided.

Despite the good mechanical properties of GRP, the high environmental impact of the glass fibre production and the difficulties in recycling and reuse have always been critical aspects. Retaining the advantages of composites, natural-fibre reinforced plastics (NFRPs) promise to be a more environmentally friendly and lightweight alternative to GRPs.

The following table presents the mechanical bending properties of structural timber (D30, according to Eurocode 5), GRP and NFRP [4]. NFRP, with a rigidity value in-between that of timber and GRP, has a bending strength more than three times higher than timber (*Fig. 2 - left*). Compared to glass-fibre reinforced plastics (GRP), NFRP properties become more interesting when regarding the values with respect to their weight-ratio (*Fig. 2 - right*). By choosing appropriate fibre volume fractions, competitive properties can be achieved.

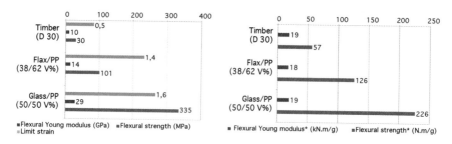

Fig. 2 Comparison of timber, NFRP and GRP mechanical properties

2.2 NFRP as Building Material

At present the demand for sustainable construction is becoming increasingly strong. One strategy for more resourceful and environmentally-friendly construction is the selection of materials and systems in terms of their life cycle assessment (LCA) and environmental impact. A further strategy is to apply the principle of building with lightweight materials and structures. In general, materials made from renewable raw materials offer strong performance in terms of sustainability and environmental impact.

NFRPs, based on rapidly renewable fibres and biobased and/or biodegradable matrices, can be used for the economically competitive and industrially scaled manufacture of unidirectional fibre-reinforced profiles by means of the pultrusion technique. NFRP profiles with a unidirectional orientation of fibres offer ideal mechanical properties for a variety of structural applications.

As mentioned above, global stability is one of the most relevant structural aspects of lightweight gridshells. Generally, it can be said that buckling loads are directly proportional to the bending stiffness of the grid profiles. Let us consider four different cross-sections (Fig. 3) each with the material properties of timber, NFRP and GRP respectively and each having the same bending stiffness (EI). The first timber cross-section corresponds to the double rectangular layer used in the Weald and Downland Museum, while the second cross-section to a classic rectangular timber profile, with the same proportions than the rectangular profiles used on the first section. Tubular cross-sections have been chosen for the composites materials, as fibre-reinforced hollow sections are feasible with pultrusion and because they offer further advantages in terms of optimisation of profile weight and cross-sectional properties.

One can see that, with less material quantity, single-layer pultruded sections are able to achieve stiffness properties equivalent to the double-layer timber system, thanks to the efficiency of hollow sections combined with the composite's higher modulus of elasticity. The NFRP tube is the lightest alternative, with a section weight corresponding to 87% and 80% that of the timber and GRP profiles respectively. As expected, the timber single-layer section is the most inefficient option, being near three times heavier than the NFRP profile.

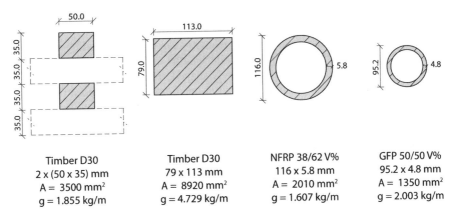

Fig. 3 Timber, NFRP and GRP cross-sections with equivalent bending stiffness

Nevertheless, the considered cross-sections also differ in the ultimate strength capacities and, consequently, in the utilisation rate induced by the gridshell shaping as well as in the ultimate bearing moment under external loads, both depending on the surface curvature of the resulting gridshell. Generally, the higher the material's mechanical strength, the later the ultimate stresses are reached and, consequently, the further the profiles are able to be bent, making possible new geometries and forms.

Depending on the gridshell geometry and the service loads that the structure will be subjected to, the considered materials become more or less appropriate. The optimal material and cross-section for a specific gridshell structure can be obtained when both aspects, global stability and the material's ultimate strength, are considered.

3 Gridshells - Shaping of Developable Grids

Gridshell geometry results from a shaping process, where an initially in-plane two-directional grid configuration is progressively elastically bent until the desired gridshell curvature is obtained. The profiles in the first direction are connected to the profiles in the second direction by pin joints which allow the distortion of the grid that is required to reproduce the target surface geometry. After the shaping process and before removing the shaping external forces, the edges of the gridshell will be fixed by rigid border beams and diagonal bracing will be added in order to maintain the desired geometry, otherwise the grid would further distort by searching an internal equilibrium, and to obtain in-plane shear bearing capacity.

Formfinding methods are required to generate grid configurations developable on specific surface geometries and to determine the internal forces and moments after shaping. Various studies have developed methodologies which can be basically classified according to the process with which the grid configurations (profile orientation and lengths) are defined. In some methods the grid is geometrically determined, while in other methods the grid results from an equilibrium of forces after applying external shaping forces to the modelled structure.

Nevertheless, the same target surface geometry can be reproduced by numerous grid configurations, differing in the profile lengths and the angle between both grid directions, once the gridshell is already shaped. The grid configurations have an important influence not only on the initial bending stresses induced during the erection process, but also on the structural behaviour of the gridshell under external loading. In this chapter, three different configurations of an anticlastic gridshell have been analysed in order to identify the structural consequences of each grid choice.

3.1 Existing Formfinding Methods

Existing formfinding methods can be classified into two groups depending on how the developable grid is defined: grids geometrically defined and grids resulting from equilibrium of forces.

Methods with a geometric definition of the grids include:

The 'Compass' Method

In 1974 Frei Otto's Institute for Lightweight Surface Structures (Institut für leichte Flächentragwerke) [5] proposed one geometric method consisting of tracing a grid with equilateral meshes on a target surface, at that time with only the help of a compass. One should start by defining on the surface two arbitrary main curves, the main profile directions, having a common intersection point. Dividing these main curves into equal segments, corresponding to the desired mesh size, one can consider the intersection point and the endings of the first adjacent segments of both main curves as three corners of the first equilateral mesh. The fourth corner can be then determined by tracing circles on the surface with a radius equal to the mesh size and centres at the two segment endings. The intersection of both circles gives the fourth mesh corner. The following meshes can be traced by using the same principle.

In 2007 M.H. Toussaint of the Delft University of Technology [6] developed a design tool based on the same principle using the three dimensional modelling CAD software Rhinoceros™. The tool consists of a script capable of automatically generating all the grid meshes on any target surface, once the main curves are given. Instead of circles, spheres are used to find the intersection points of the grid profiles. Knowing the coordinates of these points, the tool can also check the profiles' curvatures.

Although the latter method can be used to generate grids developable into specific surfaces, an additional calculation, taking into account the internal forces induced by the bending process, is required to determine the resulting gridshell geometry.

Dynamic Relaxation with Initial Plane Geometry

This method was used for the design of the Weald and Downland Museum in Sussex, Great Britain (2002) [2] and the GRP gridshell in Navier Institut, France (2005) [3]. With the dynamic relaxation method the resulting gridshell geometry in equilibrium after shaping can be calculated. This method consists on defining, the movement of the profile intersection points, modelled as nodes with a certain fictitious mass, of a pre-defined plane grid configuration, applying Newton's 2nd Law and resolving the equilibrium of forces in each node. In this equilibrium of forces, the internal forces correspond to the axial forces and bending moments generated during the construction process and the external forces to the shaping ones.

Although this method provides the effective curved gridshell geometry, the starting plane nodes configuration, and respectively the initial flat geometry, are geometrically predefined. For the Weald and Downloand Museum a squared flat grid of 49 m long and 24.2 m wide was used, while for the GRP gridshell an elliptic plane geometry was chosen. A specific target surface is difficult to be achieved with this method.

Methods with grids resulting from the application of a system of forces are:

Hanging Chain' Method

The physical *hanging chain* method was employed for the design of the first built gridshell structure, the Multihalle in Mannheim, Germany, in 1975 [1]. Neglecting the bending stiffness of the grid profiles, the *Institut für leichte Tragwerke* with Frei Otto, studied the possibility of reproducing the resulting curved gridshell geometry by means of a suspended net model with equilateral meshes. The connections between the grid profiles were modelled as hinges using rings. In order to achieve the desired surface curvatures, the number of chain members was manually modified until the final gridshell geometry was obtained. Finally, the resulting position of the nodes were redetermined by photogrammetry and, with it, the static equilibrium of the final structure was calculated using the force density method and by considering the real weight of the gridshell.

One can consider that the grid configuration results as an iterative visual process influenced by the resulting net geometry due to suspension forces. Although physically generated, equilibrium of forces help to define the final grid. With this method the effective curved gridshell geometry cannot be generated precisely, since the internal bending moments and forces, relevant for the structural equilibrium of the gridshell, are neglected.

Dynamic Relaxation with Shape Approximation

In the design methodology developed by M. Kuijvenhoven from the Delft University of Technology (2009) the grid configuration results from an approximation process, which generates a grid as close as possible to a target surface without exceeding the material's permissible stresses [7]. The curvature of the profiles is controlled by a system of transverse springs which are linked to and migrate towards the initial target grid geometry. The spring coefficients, and consequently the shaping forces, are modified iteratively so that at any point in the grid the permissible stresses are not exceeded. Once the approximate grid geometry is defined, the final gridshell geometry, modelled by a second system of interpunctual and rotational springs, is calculated by removing the action of the shaping forces. Dynamic relaxation is used here to find geometries in equilibrium, firstly, during the approximation phase and, secondly, after removing the shaping forces.

The limitations of the tool are that the grid configuration can only be modified approaching the grid vertically towards the target surface, that only one grid orientation can be considered for the shape approximation and that torsion and shear are not taken into account when calculating the resulting gridshell geometries.

Dynamic Relaxation with Application of a Vertical System of Forces

In 2009 the Navier Institute, France, proposed a second methodology where firstly the initially flat grid is set up over the target surface andsecondly a system of vertical forces, resulting from a convergence analysis, is applied to the rearranged grid so that this one is able to acquire the desired geometry [8]. The equilibrium shape of the gridshell is calculated using the dynamic relaxation method, permitting friction between the grid and the target surface. Unlike the shape approximation method, no optimisation of the post-shaping bending stresses is performed.

3.2 Finite Element Modelling with Optimisation of the Initial Bending Stresses

In the methods described above, the resultant grid geometry after formfinding generally depends on the choice of the initial grid orientation. Indeed, the same target surface can be reproduced by various grid configurations, which basically differ from one another in the resulting angle between the profiles and their lengths. In order to analyse the structural consequences of the choice of grid type and to determine the most appropriate grid configuration, a design methodology has been developed which takes into account the curvature of the bent grid profiles of a target surface and the load-bearing capacity of the resulting gridshell. This method is explained below:

Firstly, the grid configuration will be generated by means of an algorithmic calculation which tends towards minimized curvatures of the bent profiles and consequently minimized bending stresses during the shaping process. For example, a 30m long and from 14 to 15m wide anticlastic surface is considered. Three different grid configurations were generated, with acute, rectangular and obtuse angles between the grid profiles in the transverse direction, respectively. The corresponding plane geometries can be determined using three dimensional CAD.

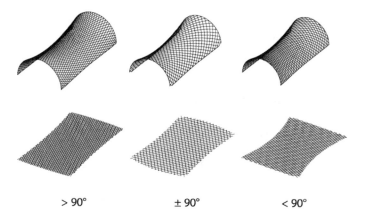

Fig. 4 Grid configurations acquiring the same anticlastic surface. Curved (above) and plane (below) geometries

Secondly, the shaping process of the optimized grid configurations is reproduced by means of finite element analysis in order to obtain resultant gridshell geometry and to evaluate the structural behaviour under external loading. One advantage of the finite elements methods is that bending, torsion as well as shear are modelled.

Minimisation of the Initial Bending Stresses

The algorithm minimizing the curvature of the profiles consists of defining energies that penalise deviation from an optimum. In our case we need to minimize three variables at once. Firstly, the curved grid configuration must remain close to the target surface, so that the positions of the profile intersection points are constrained to stay in the vicinity of the reference surface. Secondly, in order to generate a developable grid, the mesh size should be constant. And thirdly, the angle between two consecutive profile segments should be, theoretically, almost 180° in order to avoid bending stresses. This results in a linear combination of three *energies* and their corresponding coefficients.

The energies in question are typically non-convex. This means that there are many local minimizers which an algorithm might find. The key to finding a good geometry with the desired properties is now the initialization of the minimization. Once an initial guess is found, a gradient descent or the Newton method can be applied to find local minimizers of the function. In our case the initialization is found by starting with a remesh stemming from a conformal parametrization, thus having minimal geodesic curvature. From here we minimize the described function to achieve a uniform mesh size and minimal curvature of the profiles. Changing the orientation of the starting remeshed grid configuration, different solutions can be generated for the same reference surface. The corresponding plane grid geometries have been defined by means of CAD.

The advantage of this method is that by the optimisation of the initial bending stresses, the curvatures of the profiles can be modified in all directions, which helps to stay nearer to the reference surface.

Structural Analysis by Means of FEM-Modelling

The shaping process of the initially flat grid configurations and the load-bearing capacity of the resulting gridshell structures were analysed with a three dimensional, geometric non-linear finite element model using the FEM software Sofistik AG. The grid profiles and the post-shaping bracing were modelled using beam and cable elements respectively with their corresponding material and cross sectional properties. Pure geometric surface elements, without any influence on the structure stiffness, were defined in order to apply uniformly distributed external loads. The connections between the profiles in the primary and the secondary directions were modelled using coupling elements. Taking example of the clamping connections of the already constructed gridshells, kinematic constraints were defined so that only the in-plane grid distortion is permitted. An iterative equation solver (conjugate gradients) was applied.

The erection process starts with a flat grid whose longitudinal edges are constrained to the XY-plane (fixed in Z, free in X and Y). NFRP tubes of approximately 5 cm diameter have been used. Then upward nodal loads are distributed on the grid surface to induce the shaping of the structure. Once the desired geometry is obtained, the edges are fixed and the stiffening cables are added in order to maintain the desired gridshell structure and to obtain shear stability. The gridshell structure can now be loaded. The analysis was performed by applying uniformly distributed snow loads of incrementing magnitude.

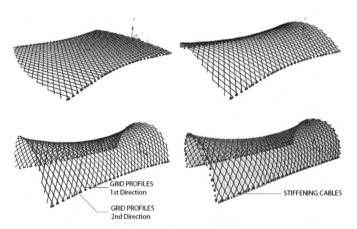

Fig. 5 Erection process of an anticlastic gridshell modelled by FEM

In the following figures (*Fig. 6*), the deformation (magnified by a factor of 20) under a snow load of 1.0 kN/m^2 are illustrated. Different deformation forms are obtained for each grid configuration.

The gridshell with predominantly acute angles between the profiles exhibits a more arch-like deformation as the gridshell contains a major number of transversal profiles which activate the arch-like structural behavior. The deformation figure consists of a single-buckle shape where the maximum deformations are concentrated in the middle of the surface. By increasing the angle between the profiles, the arch-effect decreases and a combination of arch and shell behaviors can be noticed. Indeed, the gridshell with predominantly right angles exhibits a double-buckle deformation where the central surface area sinks uniformly and the longitudinal edges, where the maximum deformations are located, buckle outwards. In the third gridshell, with predominantly obtuse angles and a major number of longitudinal profiles, the lateral edges buckle inwards stiffening the rest of the structure which sinks uniformly and lower than the first ones.

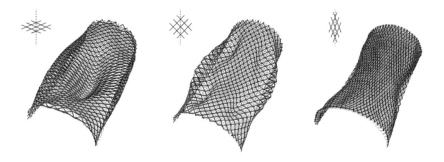

Fig. 6 Deformation shapes under snow loads

In the figure Fig. 7 (*left*) the maximum nodal displacements of the gridshell structures under increasing uniformly distributed surface loads are shown. One can see that by increasing the angle between the profiles the stiffness of the structure increases which results from the different deformation shapes the gridshells acquire. For a 1.0 load factor, the maximum nodal displacements of the gridshells with predominantly acute and right angles are about 6 and 3 times higher than the maximum deformations of the gridshell with obtuse angles.

Fig. 7 (*right*) shows the corresponding maximum von Mises stresses of the gridshells. Extreme values on the edges, due to geometric problems on the FEM-model, have been neglected. One can see that gridshells with lower angles, and consequently a major number of more curved transversal profiles, start with higher initial bending stresses. By the first loading factors, it can be observed that these initial bending stresses predominate and evolve relative slowly. At a certain load level, stresses due to external loads prevail and a relative linear evolution starts.

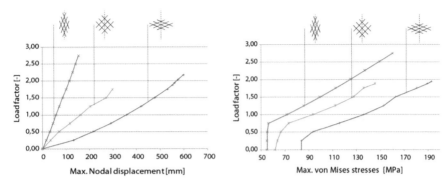

Fig. 7 Maximum nodal displacement and von Mises stresses by increasing uniformly distributed load

4 Conclusion

The final geometry of highly elastic gridshells results from a shaping process where the grid members are progressively elastically bent. In order to obtain the resulting gridshell geometry, several research studies have developed formfinding methods reproducing this erection process. Generally, these methods generate gridshells with an arbitrary orientation of the grid profiles. However, the grid orientation has an important influence not only on the post-shaping bending stresses but also on the structural behaviour of the gridshells.

The structural consequences of the grid orientation have been analysed for an anticlastic surface. Firstly, different grid configurations have been generated by means of an algorithmic calculation where the initial profiles curvatures, and consequently the initial bending stresses, tend to be minimized. Secondly, the shaping process and the load-bearing behaviour of the grids have been modelled by means of finite element methods.

The analysis show that, depending on the orientation of the grid profiles, the grid configurations adopt different deformation shapes under uniformly distributed loads, which results in variable maximum nodal displacements and von Mises stresses. Current studies analyse this structural effect for a variety of surface geometries and under diverse loading cases.

In addition, the potential of using NFRPs as environmentally-friendly alternative for gridshell structures has been studied. When regarding the bending stiffness of the profiles, NFRPs appear to be a lightweight material of choice.

References

[1] Burkhardt, B., Bächer, M., Otto, F.: Multihalle Mannheim: Dokumentation über die Planungs- und Ausführungsarbeiten an der Multihalle Mannheim. Institut für Flächentragwerke, Stuttgart (1978)

[2] Richard, H., John, R., Oliver, K., Stephen, J.: Design and construction of the Downland Gridshell. Building Research & Information 31(6), 427–454 (2003)
[3] Douthe, C., Baverel, O., Caron, J.-F.: Form-finding of a grid shell in composites materials. In: Proceedings of the International Association for Shell and Spatial Structures (IASS) Symposium (2006)
[4] Van de Velde, K., Kiekens, P.: Thermoplastic pultrusion of natural fibre reinforced composites. Composite Structures 54, 355–360 (2001)
[5] Otto, F., Schauer, E., Hennicke, J., Hasegawa, T.: Gitterschalen: Bericht über das japanisch-deutsche Forschungsprojekt S.T.T., durchgeführt von May 1971, bis May 1973, am Institut für Leichte Flächentragwerke. Seibu Construction Company / Institut für leichte Flächentragwerke / Krämer (1974)
[6] Toussaint, M.-H.: A design tool for timber gridshells. Master's thesis, Delft University of Technology (2007)
[7] Kuijvenhoven, M.: A design method for timber grid shells. Master's thesis, Delft University of Technology (2009)
[8] Bouhaya, L., Baverel, O., Caron, J.-F.: Mapping two-way continuous elastic grid on an imposed surface: Application to grid shells. In: Proceedings of the International Association for Shell and Spatial Structures (IASS) Symposium (2009)

Parametric Design and Construction Optimization of a Freeform Roof Structure

Johan Kure, Thiru Manickam, Kemo Usto, Kenn Clausen, Duoli Chen, and Alberto Pugnale

1 Introduction

The pioneering works by Poleni, Rondelet and especially Gaudì, shows how some structural principles related to the field of shape-resistant structures has been well-known for centuries. The design of structural shape has been approached differently; the more analytical way of engineering, such as the works of Dyckeroff and Widmann or the intuitive approach of engineering referring to the works of Torroja. The focus for the new structures was lightweight, large spans, functionality, efficiency and economy. This brought new developments of form-findings structures, i.e. a set of tools and strategies to find the form of 'structural minimum' - in shells where the surface is mainly stressed in the plane with compression, tension and shear. The research started with the development of experimental tools, or physical models, reaching a high point with the work of Heinz Isler [2]. However, this prosperous period has been concluded in about 20 years, since the rigid generative rules of shape-resistant structures brought to the rapid exploration of the complete family of potential shapes of shells during the 60ies.

Only with the development and introduction of computer technologies in architecture and engineering we assist to a renovated interest towards shells and shape-resistant structures in general. First, because the potential of exploring and representing any kind of complex geometry by means of NURBS extremely enhance the designer's possibilities, bringing to the development of 'non-standard' and 'free-form' investigations [3]. Second, because with computer simulation the traditional form-finding can be approached in a numerical way, reproducing the search of catenaries and minimal surfaces as happened with physical tools, but also considering the design problem as a question of 'optimization', to be performed after the main architectural choices has been already defined.

Johan Kure · Thiru Manickam · Kemo Usto · Kenn Clausen · Duoli Chen
Master's students, Master of Science in Architecture and Design, Aalborg University, Aalborg, Denmark

Alberto Pugnale
Assistant Professor, Department of Civil Engineering, Aalborg University, Aalborg, Denmark

In this paper, the design of a free-form roof structure is presented, approaching the problem in order to reduce construction costs and to define an efficient structural behaviour - see also Basso et al. [4]. The design process is supported by the use of a parametric tool, Grasshopper™, for the definition of an optimization problem related to shape-resistant structures, and then a Genetic Algorithm, Galapagos™, is used to explore/improve the shape of the 'a priori' defined structure, or better a parametric solution domain of tentative structures. Finally, a scripting interface between the CAD software, Rhinoceros™, and the FEM solver, Autodesk ROBOT™, is described as a rapid way to check and refine the structural behaviour of the overall roof.

2 The Project

The optimization procedure described in this paper has been developed, starting from a design proposal for the new Historical Museum of North Jutland, in Denmark, as reference project. The program for this new museum has been defined both as a closed design competition and a design studio for master's students in Engineering, Architecture & Design at Aalborg University, during the fall semester 2010. The site of the project is located in the landscape near Fyrkat, Denmark. The main design issues to be addressed have been related to the topics of tectonic and Nordic architecture. The definition of a Viking Museum therefore focused on construction, structural and material aspects, as well as the perception of architectural spaces integrated into the landscape.

Therefore, the building has been conceived as a free-form ruin-like heavy concrete base, directly anchored into the landscape. The roof is in contrast a light free-form shell resting on top of the base, and is made up of timber panels assembled in a triangular faceted form. This would be perceived as a cave-like room from the inside, emphasized by means of large timber columns, which are cutting through the geometry as space defining elements (Fig. 2.1 & Fig 2.2).

Fig. 2.1 & 2.2 Exterior and interior rendering

3 Parametric Definition of the Morphogenetic Problem

3.1 Mesh

The conceptual idea for the project gives the possibility to implement computational techniques. By considering structure, construction and assembly, it was possible to investigate and develop the free-form roof shell, with the means of a morphogenetic optimization procedure.

For this reason, the architectural element of the roof is initially defined in parametric terms with Grasshopper™ in order to investigate design variables and constraints. First, the reference geometry is defined by three guide curves, lofted to create a NURBS surface. Second, a Delaunay triangulation algorithm is used to construct a triangular mesh on this surface starting from a set of points in three-dimensional space. The solution domain can be finally explored varying the geometry of the three guide curves used to generate the reference surface, and the position of a set of points, a 'point cloud', placed in the plan is projected on the reference surface for the definition of the triangular mesh (Fig. 3.1).

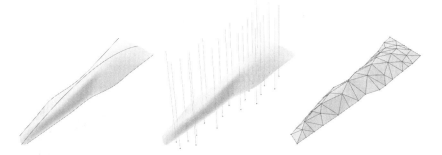

Fig. 3.1 From smooth surface to Delaunay mesh

3.2 Component

Each triangular element results from this first parametric definition of the roof and is used as a geometric boundary for the design of the final timber structural panels. They are studied in order to reduce manufacturing and assembly complexity and parametric adjustments are made according to the overall shape. Applying a recursive subdivision algorithm generates a structural element for each roof surface, following this procedure: First, each triangle is divided into four sub-triangles. Second, the respective edges and centroids are connected. It should be underlined that such a subdivision method uses the circumcircle centroid of each triangle, as well as the midpoint on each triangle segment to construct the components. The reason for using the circumcircle centroid instead of the area centroid of the triangle is to avoid joints with three-dimensional rotation. By keeping the joints two-dimensional it is possible to fabricate the elements on a 3-axis CNC milling machine.

In such a parametric definition of the roof structure a geometrical issue arises when a triangular component has obtuse angles, i.e. the circumcircle centre of a triangle does not lie inside the triangle. In this geometrical condition, the circumcircle centre will land outside the triangle, causing the subdivision algorithm to give an output that is not suitable for structural purposes, because of the non-perpendicular meetings (Fig. 3.2).

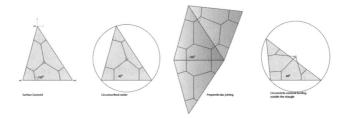

Fig. 3.2 The perpendicular meeting achieved by using circumcircle, and the problem when the corner angle exceeds 90 degree

4 The Optimization Procedure

Solving this geometry and construction problem for each component in three-dimensional space requires the use of an optimization technique. In this case, a Genetic Algorithm [5] is chosen for a set of reasons. First, it allows a wide exploration of the solution domain by means of a metaheuristic search method, which can be easily followed by the designer and give inspiration and direct feedback during the optimization process. In this situation the goal is not to reach the optimal solution, but to define a sub-optimal solution to be considered as the best compromise after an in-depth evaluation of design criteria. Second, because GAs does not need the definition of an initial design proposal, i.e. a first tentative solution, but just a 'solution domain'. For the designer that means the formulation of a problem in parametric terms, their respective relations and range of variability. Third, because a new Genetic Algorithm called Galapagos™ has recently been developed and introduced as a tool inside Grasshopper™, providing a direct link to the parametric definition of the problem worth investigating.

4.1 GALAPAGOS GA

Galapagos™ [6] is a user-friendly GA, it allows for a direct definition of design variables and solution domain by means of Grasshopper sliders, and the definition of an objective function, or in GA technical vocabulary a 'fitness function' by means of a floating number, which can be minimized or maximized. No information is provided about types of selection. Crossover and mutation operators are used, and little control is given the user in relation to choice of the number of individuals per population, number of maximum generations i.e. iterations of the algorithm, and percentage of application of genetic operators (Fig. 4.1).

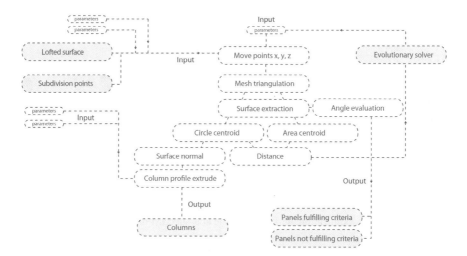

Fig. 4.1 Flow-chart of the optimization algorithm

5 Fitness Function and Solution Domain

The parametric definition of the roof with triangulation is used to define a solution domain for the Genetic Algorithm, Galapagos™. By decomposing each corner point, the triangles are then given a range of freedom, or variation in three-dimensional space. Such a limit is defined by the designer as a balance between desired form and degree of optimization. This allows Galapagos to modify and evaluate the overall shape according to the fitness criteria.

The fitness function has to be described in order to allow the algorithm to evaluate if each triangle fulfills the criteria for its angles. The fitness is the minimum distance between circumcircle centroid and area centroid, thus minimizing obtuse angles and avoiding the circumcircle centroid falling outside the triangle boundary. The distance is evaluated, for each modification on the point coordinates, and if it exceeds a given distance the solution is given a penalty by multiplying the distance exponentially. This guides Galapagos™ in selecting the fittest populations to further breed on and the population of fit individuals goes towards the best possible solution. (Fig. 5.1)

Fig. 5.1 Process of optimization, dark facets is the non-successful triangles

6 Results and Further Developments

The optimization focused on rationalizing the triangulation of a pre-established roof surface, with the constraint that is should respect the building plan as its boundary. The geometric optimization done with Galapagos™ allowed the archiving of a structure where 80 percent of the triangles succeeded in not having any corner angles above 85 degrees. This is a product of a very narrow space of freedom given to Galapagos™.

The advantage of a GA compared to a conventional linear way of solving lies in the variety of solutions generated. To reach a good result, it needs the possibility to operate on a broad range of solutions, but this also requires a considerable amount of time. From a design point of view it implies the possibility of investigating a broader range of informed design solutions and the possibility for discovering new and interesting solutions to a design problem. A potential for the GA is that it could be used as an active tool to explore new design solutions e.g. informing the form and plan of the building (Fig. 6.1).

The use of these in the design process relies on the designer's ability to set up the right solution space and fitness criteria. A way of investigating these possibilities could be through the setup of a solution space with a large flexibility in the modification of points, but also larger steps in between solutions enabling the solver to give a feedback on the widest possible solution space.

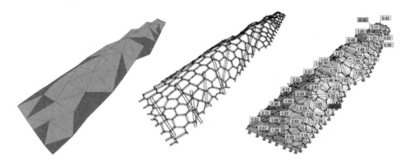

Fig. 6.1 Results of FEM structural analysis diagram

The optimization focused primarily on fabrication issues, but was also simultaneously during the design process evaluated for its structural properties. The procedure with GA showed that small variations in the node placements of the triangle corner could significantly reduce stresses in the overall structure. The Structural evaluation has been done with the finite element program Robot Analysis™ from Autodesk.

The open programming interface in the Robot API made it possible to program a direct link from Rhino to Robot making the Structural Calculations part of an Iterative design procedure enabling quick evaluations on the changes in the Structure. Due to time limitations, the investigation was solely done for uniform dimensioned structural bars (width and height).

A planned further development would involve letting Galapagos inform the dimensions according to obtained stresses in each bar member allowing a material optimization as part of the architectural expression.

7 Conclusion

As shown by this case study, the procedures of optimization by means of Genetic Algorithms and finite element analysis can be used in a parametric workflow to create an approach of integrated architectural design. Given the easy access to a GA in the form of the Galapagos solver in Grasshopper the designer now has easy access to a powerful tool, which can be used in an informed design exploration. The method can be directly implemented with the parametric model early in the process and requires most importantly the deliberate structure of the model.

References

[1] Torroja, E.: Philosophy of structures, 1st edn. University of California Press, Berkeley (1958)
[2] Isler, H.: New Shapes for Shells, IASS Bulletin, no.8-C3, Madrid (1959)
[3] Migayrou, F. (ed.): Architectures non standard. Centre Pompidou, Paris (2003)
[4] Basso, P., Del Grosso, A., Pugnale, A., Sassone, M.: Computational morphogenesis in architecture: cost optimization of free form grid shells. Journal of the International Association for Shell and Spatial Structures 50(162), 143–150 (2009)
[5] Mitchell, M.: An Introduction to Genetic Algorithms. The MIT Press, Cambridge (1998)
[6] Rutten, D.: Evolutionary Principles applied to Problem Solving. Grasshopper3D (2010), http://www.grasshopper3d.com/profiles/blogs/evolutionary-principles (accessed June 13, 2011)
[7] Fisher, A., Sharma, S.: Exploiting Autodesk Robot Structural Analysis Professional API for Structural Optimization. Autodesk (2010)

Curved Bridge Design

Lorenz Lachauer and Toni Kotnik

Abstract. This paper presents a novel approach for the interactive design of linear structures in space. A method is introduced, that provides a maximum of formal freedom in the design of funicular, hence efficient, structures. For a given deck geometry, defined by a design driving NURBS curve via parametric modeling techniques, a tailored relaxation routine allows for controlled, real-time form-finding of the spatial funicular. Subsequently, the equilibrium of the deck is constructed using techniques from graphic statics, combined with a least-square optimization technique. Finally, the method is applied to a design example.

1 Introduction

The role of computation in architecture has undergone a shift in the last few years: from the promise of novel, unlimited freedom in form, enabled through digital design and fabrication processes, to a more critical attitude, questioning the complexity and cost of this freedom [15]. This change of focus is one reason for recent ambitions of the integration of structural and fabrication constraints into architectural design practice. [14]. An engineering practice is emerging, that blurs the role of the engineer in the design process between problem-solver, and creator of ideas and design concepts [5].

One basis for this new model of collaboration between architects and engineers is an upcoming set of shared computational concepts and interfaces as common language in the design practice [16]. The integration of structural constraints with fabrication data in a tailored, real-time design environment has been described for the case of a digital hanging chain modeler [6]. Recently, a method for the setup of an associative model of an efficient structure for a freeform roof, using techniques from graphic statics, has been reported [7].

Lorenz Lachauer · Toni Kotnik
Chair of Structural Design, ETH Zurich, Switzerland

In this case study, a new computational strategy is introduced, that allows for real-time design of spatial bridges in early design stages, between the conflicting priorities of exciting spatial expression and efficient material use. Using concepts derived from form-finding methods for cable nets [1], and graphic statics [17, 13], a set of funicular polygons is generated, that are in equilibrium with the dead-load of a given deck geometry. These structures are highly efficient for the design load, due to the funicular form of the equilibrium solution.

The structure of this paper is as follows: first, the basic concepts of the technique are briefly summarized. Second, the design method and its implementation in a parametric setup are described. Finally, the design of a footbridge is shown, demonstrating the effectiveness of this approach.

2 Concepts

2.1 Equilibrium Solutions

The approach presented here is suited for form-finding in early design stages, on the basis of dead-load. The method is considering the equilibrium of axial forces in all structural members, while neglecting material stiffness and deflections. This is valid based on the assumption of the rigid-plastic material model and the lower bound theorem of the theory of plasticity [4]. In later stages of the design process, after member dimensions have been assigned to the structural axes, further structural analysis has to ensure that the deflections stay small, and that no elements are in the danger of buckling. Since funicular structures are instable systems, additional stiffening means against collapse in the case of asymmetric load have to be introduced.

2.2 Curved Bridges

In what follows we will exemplify the method by dealing with the design of a typology of bridges, consisting of a curved deck in plan, and a funicular structure, a cable or an arch, supporting the deck. Prominent build examples of this typology are the suspension bridge in Kehlheim, Germany, and the arch bridge in Oberhausen-Rippenhorst, Germany, both designed by Schlaich, Bergermann and Partner [2], and the Campo Volantin footbridge in Bilbao, Spain by Calatrava [3]. For the general case of a bridge in equilibrium, with its deck following an arbitrary space curve, it is always possible to find two funiculars, balancing the dead-load of the deck [9]. For this paper, a specific case is discussed: the geometry of the deck is arbitrarily curved, but lying within a horizontal plane. In Fig. 1, the main concept for the form-finding of the funicular is described for a suspension bridge: the goal is to achieve a

Curved Bridge Design

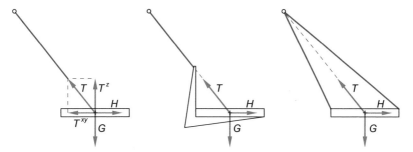

Fig. 1 Equilibrium of a vertical section of a curved suspension bridge, for three different cases of construction: the cable is fixed in the centroid of the deck (left), the cable is fixed at an additional element working in bending (middle), the cable is split into two cables fixed at the ends of the section (left) (after [18])

force T in each cable, in order that the vertical component T^z balances the dead-load G of the deck G. The remaining component T^{xy} has to be balanced by the force H within the horizontal plane of the bridge deck [18].

3 Design Method

3.1 Form-Finding of the Spatial Funicular

In this section, a computational method for the generation of the funicular polygon in space is described. It is based on a form-finding technique for tension structures using dynamic relaxation (DR) [1]. In this paper, DR is not explained to full extend, only the differences to Barnes's method are identified.

The anchor points $S_1 \ldots S_n$, on the axis of the bridge deck are equally spaced, so one can assume the dead-load $G_i = G$ for all i. The nodes of the funicular at time t are named $X_0^t \ldots X_{n+1}^t$, the factor r controls the rise of the funicular (see Fig. 2). The supports X_0, X_{n+1}, are input parameters.

As illustrated in Fig. 1, the funicular has to fulfill: $-T_i^z = G_i$ for all connecting elements between funicular and deck. In order to reach this condition, the residual force at node X_i^t, for $0 < i < n+1$, is considered as $\vec{R_i^t} = \vec{F_i^t} + \vec{F_{i+1}^t} + \vec{C_i^t}$. Similar to DR, the forces in the funicular at node X_i^t are determined in relation to their initial length at time $t = 0$: $\vec{F_i^t} = \overrightarrow{X_i^t X_{i-1}^t}/\|\overrightarrow{X_i^0 X_{i-1}^0}\|$ and $\vec{F_{i+1}^t} = \overrightarrow{X_i^t X_{i+1}^t}/\|\overrightarrow{X_i^0 X_{i+1}^0}\|$. The difference to DR is the definition of the forces C in the connecting elements between deck and funicular:

$$\vec{C_i} = \frac{\vec{Q}}{Q^z} \cdot r, \text{ with } \vec{Q} = \overrightarrow{X_i^t S_i} \tag{1}$$

Equation (1) ensures that the magnitudes of the vertical force components of all C_i^t are r. The solving procedure for the solution X^* is straightforward using DR. The forces T are finally scaled by the factor G/r in order to balance the dead-load vertically: $\vec{T_i} = \vec{C_i^*}\frac{G}{r} = \frac{\vec{Q}}{Q^z}G$.

3.2 Implementation of the Interactive Tool

A tailored interactive tool for the NURBS modeling software *Rhinoceros 4.0* [10] is developed, within the existing associative modeling environment *Grasshopper* [11]. This tool allows for the interactive design of spatial funicular polygons, based on the input parameters described in Fig. 3. The tool consists of a custom Grasshopper component, written in VB.NET, which iteratively solves for the funicular, using DR with kinetic damping. In each step, the residuals R_i^t are updated as described in the preceding section. The residuals are used to calculate nodal velocities and resulting node locations [1]. The number of iterations is adjustable, in order to control the solving speed.

For models with a high number of segments, it is useful to first use a low number of iterations in order to maintain real-time behavior of the tool. As soon as the funicular is near to the desired form, the designer can increase the number of iterations in order to get a more accurate solution. The plot of the convergence graph helps to identify instabilities in the solving process; change of the nodal mass allows controlling the convergence speed. For the example shown in Fig. 4, using a appropriate nodal mass setting, the funicular updates for 20 segments in real time, 55ms, after 200 iterations. A highly accurate solutions after 2000 iterations with a residual that is smaller than 1e-11 of the dead-load is obtained after 5000ms, using an Intel Core Duo processor with 2.8GHz. Both results are visually very close, the maximal distance between the two funiculars is roughly 0,1% of its length.

Fig. 2 Initial state of the form-finding process (left), the funicular as tension cable for a rise $r > 0$ (middle), the funicular as arch, acting in compression for a rise $r < 0$ (right)

Curved Bridge Design

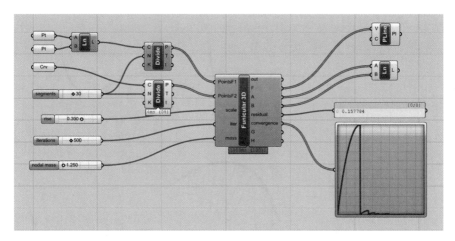

Fig. 3 A custom component for Grasshopper, generating the three-dimensional funicular. The input parameters on the left side are: the two *supports* of the funicular as points, the *deck* as NURBS curve, the number of *segments*, the rise, the number of *iterations*, and the *nodal mass*. On the right side, the output parameters are: the *spatial funicular* as polyline, the *cables* as lines, the sum of the magnitudes of the *residual forces R*, and the *convergence* graph. Additionally, the forces G and H that are sufficient for balancing the forces in the cables are calculated.

3.3 Horizontal Equilibrium of the Deck

The previous sections described a method that allows for the design of a funicular in compression or tension that balances the dead-load of the bridge deck. According to Fig. 1, there is still an unbalanced horizontal residual force, the component $T_i^{xy} = H_i$. All these residuals H are lying in one plane, namely the same as the bridge deck. Hence, either a simple funicular, or the deck in combination with an additional funicular is sufficient, in order to balance the horizontal components [19] (see Fig. 5). Previously, a set of parametric tools has been described, that allows for the interactive generation of these planar funiculars. Furthermore, it is described how to combine these tools using a generic minimization routine that is a build-in component of *Grasshopper*, in order to find the closest funicular to a given curve using a least-square approach [8].

4 Results

The method has been applied for the design of a fictional footbridge project in steel (Fig. 6). First, the spatial funicular arch has been generated for the chosen deck geometry, spanning the hypothetical 50m wide river. As a next step, the constructive concept based on two suspension cables supporting the bridge deck in each vertical section, as shown above (Fig. 1, right), has been implemented with parametric solid

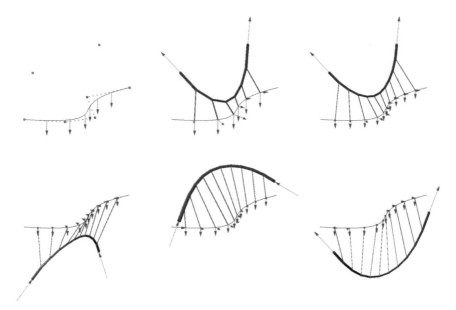

Fig. 4 Variations of the spatial funicular: points as supports of the funicular and a NURBS curve as deck axis, divided in seven segments (above left), a possible funicular for a positive rise (above middle) a funicular for the same supports, deck curve, and rise, with eleven divisions (above right). Three different funiculars for the given deck geometry, with varying supports positions and rises (below). The axial forces are rendered as cylinders, the diameter is proportional to the magnitude of the force. The external forces G and H sufficient to balance the deck, are rendered as arrows.

Fig. 5 Horizontal equilibrium of the deck: a funicular with the forces G and H, that are able to balance the deck (left), the horizontal forces H balanced by an arch in compression (middle), the forces H balanced by a combination of pretensioned deck and a horizontal funicular in compression (right)

volumes. Finally, a structural scheme similar to the one depicted in Fig. 5 right has been introduced to balance the horizontal forces in the deck. The central axis, a tube under compression, together with an additional tension cable lying in the plane of the deck, is forming a horizontal truss. A major drawback of the form-finding

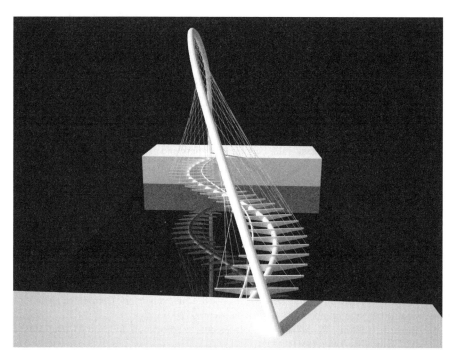

Fig. 6 Main load-bearing elements of a curved, cable-stayed pedestrian bridge

method is the lack of control of the direction of the cables in plan. The structure is in equilibrium for dead-load, but it is sensitive to asymmetric loads, additional bracings in the plane of the deck could be introduced for stiffening. The next step in the design process is the export of the structural axes to a FEM software, in order to analyze the bridges behavior for asymmetric and moving load, its vibration and deflection characteristics. Based on this analysis, final member dimensions are assigned. Furthermore, this geometry can be used as starting point for a non-linear structural optimization approach.

5 Conclusion and Future Work

This case study investigates an engineering concept that allows for the generation of funicular, hence efficient, structural forms. Through the implementation of the method in an interactive parametric setup, the abstract concept becomes tangible. Visualization and real-time feedback lead to an intuitive understanding of complicated structural relations and opens up new formal resources that are inherent in the concept. This reduces the gap between engineering and design for this specific task, and allows for the creation of structurally efficient *and* spatially exciting structures, with computational means. Although the approach is geometrically flexible, it is limited to a narrow and specific field of design. Furthermore, the design method allows for form-finding in early design stages, but requires still a lot of engineering knowledge when it comes to later design stages. Future work will deal with overcoming of the typological limitations in this approach. As a first step, the fully 3-dimensional deck curve not restricted to planar cases will be processed, by balancing the horizontal forces with non-planar funiculars. Beyond this, a general concept for the combination of multiple spatial funiculars, compression and tension, with an arbitrary connectivity will be addressed.

Acknowledgements. We thank Julie Felkner for patient and careful editing.

References

1. Barnes, M.R.: Form Finding and Analysis of Tension Structures by Dynamic Relaxation. Int. J. Space Struct. 14, 89–104 (1999), doi:10.1260/0266351991494722
2. Baus, U., Schlaich, M.: Footbridges, pp. 105–121. Birkhäuser, Basel (2007)
3. Frampton, K., Webster, A.C., Tischhauser, A.: Calatrava Bridges, pp. 206–213. Birkhäuser, Basel (1996)
4. Heyman, J.: Basic Structural Theory, pp. 26–38. University Press, Cambridge (2008)
5. Kara, H.: Design Engineering. Barcelona, Actar 9 (2008)
6. Kilian, A.: Linking digital hanging chain models to fabrication. In: Proc AIA/ACADIA, vol. 23, pp. 110–125 (2004)
7. Lachauer, L., Kotnik, T.: Geometry of Structural Form. In: Ceccato, C., Hesselgren, L., Pauly, M., Pottmann, H., Wallner, J. (eds.) Advances in Architectural Geometry, pp. 193–203. Springer, Heidelberg (2010)
8. Lachauer, L., Junhjohann, H., Kotnik, T.: Interactive Parametric Tools for Structural Design. In: Proc IABSE-IASS (2011); accepted for publication
9. Laffranchi, M.: Zur Konzeption gekrmmter Brcken, pp. 23–30. Dissertation, ETH Zurich (1999)
10. McNeel, R.: Rhinoceros NURBS modeling for Windows. Computer software (2011a), http://www.rhino3d.com/
11. McNeel, R.: Grasshopper generative modeling for Rhino. Computer software (2011b), http://www.grasshopper3d.com/
12. Muttoni, A., Schwartz, J., Thrlimann, B.: Design of Concrete Structures with Stress fields. Birkhäuser, Basel (1996)
13. Muttoni, A.: The Art of Structures. EPFL Press, Lausanne (2011)
14. Oxman, R., Oxman, R.: Introduction. In: Oxman, R., Oxman, R. (eds.) The New Structuralism: Design, Engineering and Architectural Technologies. John Wiley, Chichester (2010)
15. Picon, A.: Digital Culture in Architecture, pp. 8–14. Birkhäuser, Basel (2010)
16. Rappaport, N.: Support and Resist: Structural Engineering and Design Innovation, pp. 7–11. Monacelli Press, New York (2007)
17. Schwartz, J.: Tragwerksentwurf I. Lecture Notes, ETH Zurich (2009)
18. Strasky, J.: Stress ribbon and cable-supported pedestrian bridges, pp. 155–160. Thomas Telford, London (2005)
19. Zalewski, W., Allen, E.: Shaping Structures, pp. 377–400. John Wiley, New York (1998)

Linear Folded (Parallel) Stripe(s)

Rupert Maleczek

Abstract. This paper presents the research to find a computational method for creating freeform structures consisting of simple linear folded (parallel) stripes [Fig. 1]. The author developed a geometric algorithm that enables a structuralisation from single curved to double curved surfaces. The term structuralization stands here for the approximation of a given surface with linear folded stripes. The algorithm produces a series of stripes that form an irregular hexagonal honeycomb structure from a given surface. These stripes are rectangular in unrolled condition, and get no torsion when folded.

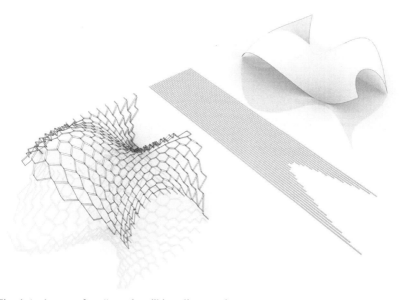

Fig. 1 A given surface "translated" into linear stripes

Rupert Maleczek
Institute for Structure and Design, University of Innsbruck, Austria

1 Why Using the Linear Folded Stripe(s)?

The "technical human made nature" is producing most materials either in rectangular plates, linear strings with multiple section profiles or in very small pieces. The idea behind the project is, to find possible ways to build freeform structures with regards to material efficiency, prefabrication and mobility. One possible approach to be efficient in the use of plate material by minimizing the produced off-cuts, is to work with rectangular stripes.

The genotype of the "linear folded stripe" describes a rectangular linear stripe with one or more folding edges [Fig.2](Maleczek 2011).

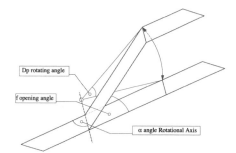

Fig. 2 A simple stripe and how to create a hexagonal structure

In order to create a hexagonal structure two stripes with 4 folds each have to be used [Fig.3](Kudless et al. 2008).

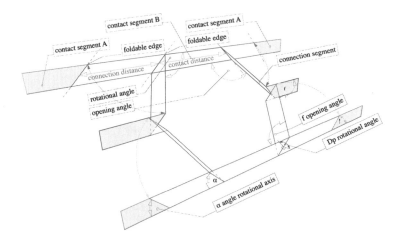

Fig. 3 How to create a hexagonal structure

After an assembly of several hexagonal cells, a reticular structure will be created. In a cellular structure every stripe consists of alternating contact- and connection segments (Maleczek 2011).

1.1 The Parallel Stripe

If the contact segments of a stripe are located in parallel planes and have parallel main directions, the system can be described as parallel stripes. Between two parallel contact segments the connection segment is always a parallelogram. In other words, the angle of the rotational axis, the opening angle and the rotational angle are equal on both ends.

A very simple example is a rectangular planar surface structuralized with a hexagonal grid. If one main axis of the plain in which the contact stripes are located is equal to the surface normal of the planar surface, then the angle α is 90 degrees, the angles Dp and f can vary from segment to segment. In any other case the opening angles and rotational axis will change according to the angle with the surface normal [Fig. 4].

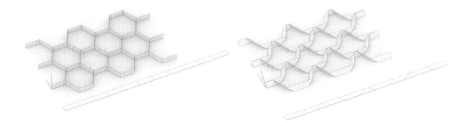

Fig. 4 A planar surface structuralized with parallel stripes in Surface Normal (left) and with stripes 45-degree angle to the Surface Normal (right)

1.2 Generation

There are two possibilities to create a connection segment within two contact segments: a mathematical and a geometrical one [Fig. 5]. In any case, the parallel contact stripes should be defined in space (e.g. on a given surface). In the se-cond step the connection segment has to be defined.

In opposition to the freely oriented stripe system, the parallel stripe system allows to define the position of the generating line of the connection segment by drawing a simple line between the contact segments. Every line between two contact segments will generate a solution.

In the geometrical approach, a rotating cylinder with the height of the stripe is intersected with the distanced edges of the contact segments, to find the right angle for the rotating axes.

In the mathematical approach, the angles of the folding axes are directly calculated. A simple formula proposed by Thierry Berthomier with the use of spherical trigonometry gives the relationship between the three main angles of a stripe: the angle alpha between the folding axe and the border of the stripe, the angle f between the borders of the stripe, and the dihedral angle D between the two planes separated by the fold.

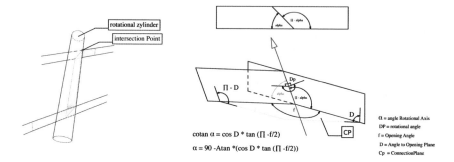

Fig. 5 Creation of a stripe: geometrical (left) mathematical after Th. Berthomier (right)

Creating a hexagonal grid with the open linear folded stripes, the algorithm creates two contact stripes that have the same base length to calculate the connection segments in between. If the following contact segments do not have a mirrored position in space, the angle of the rotating axis will not be the same. This produces an irregular node at the endpoints of each element. Here the location of the generation line within the stripe becomes a key issue to the stripe generation[Fig. 6].

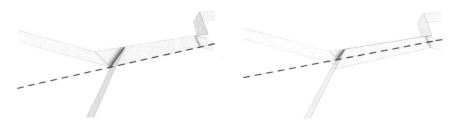

Fig. 6 Position of the generating line with two possible positions

1.3 Surface Approximations

The stripe presented here belongs to the family of post defined open stripes. In order to create a hexagonal pattern on a surface, in the first step the different contact segments have to be placed on the surface. Caused by the fact, that all contact segments are parallel, these segments can only "touch" the surface in a Point

[Fig. 7 left]. An exception are lateral surfaces, where the direction of the generating straight line is parallel to the main direction of the stripe [Fig. 7 right]. Normally the generating line of the connection segment does not lie in the surface.

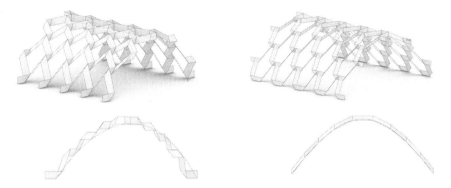

Fig. 7 A given single curved surface with two possible stripe patterns in main UV direction. The section below shows the approximation to the surface.

As one standard configuration for surfaces that will be structuralized, the surface tangents should be smaller then the direction vector of the (height + PI). If the surface curvature exceeds this limit, a special detail for the stripe assembly has to be developed (e.g. the contact segments have different lengths in the turning zone) [Fig. 8].

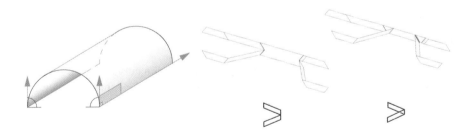

Fig. 8 The critical point of a surface and two possible solutions for a vertical stitch

The more the surface normal of the given surface tends towards the surface normal of the contact segments, the more difficult it becomes to structuralize the surface [Fig. 6.2, left]

As general rule can be proposed: if the hexagonal grid viewed projected normal to the direction vector of the stripe, can be seen without any crossing lines, the structuralisation of the surface should be possible.

There are many possibilities to distribute the starting points on a given surface. For all possibilities the direction and the length of the contact segments is one of the key factors to both the distribution of the starting points, and the form of the cells. As longer the contact segments become as more the cells transform form honeycomb like hexagons to z-shaped hexagons. The minimum length of each contact segment is defined by the height and its angles of the limiting rotational axis [Fig. 9].

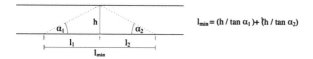

$l_{min} = (h / \tan \alpha_1) + (h / \tan \alpha_2)$

Fig. 9 Minimal length for contact stripes

The formula for the minimal length is only working, when the angle Dp < 90°. Otherwise the distance and position of the following contact stripes has to be taken into the formula.

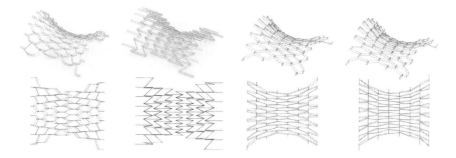

Fig. 10 A surface structuralized with different directions and contact stripe lengths

As the system of parallel stripes can not follow the main curvature of a surface, surfaces that are bend seen normal to the main direction of the contact segment, it will be necessary to generate "stitched" hexagonal patterns [Fig. 11].

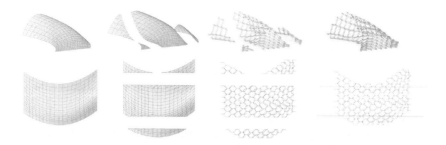

Fig. 11 A possible solution to stitch the pattern

1.4 Structural Behaviour

The actual research shows, that the form of the cells has an enormous correlation with the stability of the system. Each hexagonal cell should have angles Dp > 90°. Models and Prototypes have shown that if one or more angles Dp < 90°, the cell starts to become weak. Not only from an aesthetic view, but also from a structural view each angle in a cell should have a tendency towards 120°. As shown in Figure 5 the knot configuration is essential for the forces in the element. Here a solution with the same connection stripe length in the system reduced the bending forces in the element itself compared to other configurations.

In grid configurations, where the main direction of the stripes is located in the main force direction comparable to a wood system, the advantages of an arch like system can be used.

The actual research with models and prototypes show, that there are some correlations to the classical grid shell system. In Order to create stable freeform structures, the structure needs a fixed boundary condition, and a system to fix the angles of the cells brings a big effort to the stability of the "shell". This fixed boundary can also be made from linear folded stripes that connect every second connection segment of the limiting stripes.

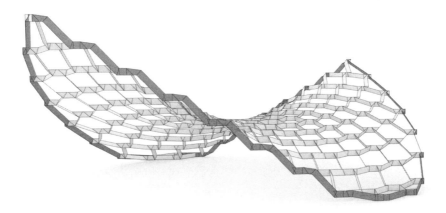

Fig. 12 A possible fixed beam boundary condition made of linear folded stripes

2 Conclusions

The research presented here shows the enormous potential of linear folded parallel stripes, to create freeform structures. Advantages and problems of this system with possible solutions to the defined problems are clearly explained.

The presented algorithm can be seen as a contribution to the wide field of geometric folding algorithms (Demaine E, O'Rourke J,(2007))

As the parallel stripe can be seen as one aspect of linear folded stripe systems, this paper can be seen as a small piece of a deeper research in this field. Ongoing research of the author will not only show and investigate different stripe topologies, (e.g.: freely oriented post defined stripes, pre defined open linear folded stripes...) but also work in different fields of the stripe (e.g.: mobility, materiality...)

References

Maleczek, R.: Linear folded stripe(s). In: Algode Conference, Tokyo Japan (2011); Publication delayed because of Fukushima Desaster...

Kudless, A., Hensel, M., Menges, A., Weinstock, M.: Honigwabenstrukturen. Arch+ 188, 58–59 (2008)

Demaine, E., O'Rourke, J.: Geometric folding algorithms, Cambridge, New York (2007)

Delarue, J.M.: Constructions Plisses – Rapport Final, Ecole d'Architeture Paris Villemin, France (1981)

The Potential of Scripting Interfaces for Form and Performance Systemic Co-design

Julien Nembrini, Steffen Samberger, André Sternitzke, and Guillaume Labelle

Abstract. This paper discusses the advantages of using a coding interface both to describe form and run performance simulations in the context of architectural design. It advocates for combining recent interest in the design community for parametric scripting with available expert-level Building Performance Simulations (BPS) to enable designers to encompass performance-related design questions at the early design stage. Pitfalls when considering non-standard solutions and the potential of the approach to circumvent such difficulties are exemplified through a housing building case study, emphasizing the under-evaluated role of the analysis tool in steering design decisions. The contribution outlines how providing designers with exploratory tools allows to consider sustainable construction in a systemic manner.

1 Introduction

When starting developing a design proposal, a common approach among architects is to explore in parallel several variants by gradually refining them until one takes precedence. Typically, elements of the design such as context, overall form or typology are firstly considered to explore their association. To allow for modification, they are at the beginning partly defined and become more and more set as the design process unfolds and decisions bind interrelations between them.

These larger scale design elements are known to dramatically influence the performance of the projected built form [3]. Finer design details such as window openings, shading elements or HVAC strategies are typicallly left for later design stages, when form and typology has been agreed by the client. However, such details also greatly impact on performance and are in strong interaction with form or context.

Julien Nembrini · Steffen Samberger · André Sternitzke
Structural Design and Technology Chair, Universität der Künste, Berlin

G. Labelle
Media and Design Lab, École Polytechnique Fédérale de Lausanne

Fig. 1 Urban infill housing case study, *(left)* urban context typical to Berlin, *(middle)* facade may include oriels, *(right)* typology with buffer space acting as usable thermal insulation

To tackle such interactions, designers tend to rely on rules of thumb such as "use difference of sun elevation to gain energy in winter and avoid overheating in summer", which represent indeed good heuristics but cannot express the complexity of material, heat and light interactions of a real design context. The aim for better performing built form thus calls one to consider the whole building system including relevant elements at all levels of detail.

To this end, non-expert usage of computer models for early design performance assessment has been acclaimed to achieve higher levels of sustainability through energy efficiency [1]. There has been several proposals to achieve such a task, which can be grouped into two categories [3]: simplified tools or subset of expert-level tools. By far the most popular, the first approach tends to over-simplify geometry and/or uses models considered too unreliable by building systems engineers [9]. Examples of the second notably restrict usage to implemented functionality, hindering customization and lack possibility to tune accuracy to address demanding design questions such as natural ventilation potential.

In parallel, architectural design research and practice has lately witnessed the growing use of *scripting* — using computer (visual) code instructions to define architectural form — and *parametric design* — maintaining dynamic links between parameters and their use in form definition for real-time continuous modifications. Despite being associated with ambitious form, these techniques have strong potential for generating and exploring early design variants: using parametric scripting, designers are able to automate geometric description and modification of architectural form. By extending this paradigm to Building Performance Simulation (BPS), it becomes possible for non-experts to generate sufficient design details to run full-featured BPS tests at an early stage, with parametric modifications further enabling design alternatives comparisons. The research presented here demonstrates, through the development of a parametric scripting performance framework, how this designers' interest to define form through code provides an innovative context to include complex systemic performance assessment at early design stage.

Fig. 2 Schematic view of the scripting/simulation framework

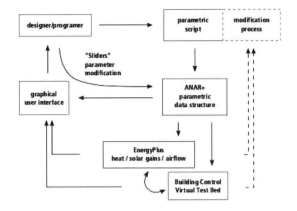

2 Approach

The research presented here proposes a specific approach in generating form exclusively through code instructions. Starting with hand sketches, designers write shape-describing code by creating and assembling points into faces and objects through geometrical transformations. This results in decoupling form definition from form representation and forces designers to "think before modeling". By compiling the code to represent the form and applying changes on the code rather than on the representation, designers are engaged in a *reflexive process* in the sense of [8]. From experience gained in design studios, the approach encourages designers to question, make precise and simplify their intentions by themselves [4].

This context naturally lends itself to using code instructions to define material properties and detail input needed by expert-level BPS, allowing for the involvement of full-featured validated BPS at the onset and providing performance feedback as soon as a form is defined. Although limited in absolute terms by users' lack of expertise, such testing provides early insight in key parameters influence over performance. Potential incoherence of result encourages further *reflection-in-action* [8] and leads the designer towards addressing under-defined areas of the design.

As a case study, a housing building in an urban infill context is considered, comparing performances of a standard typology with an innovative functional layout in interplay with environmental conditions (Fig. 1). Aiming for passive solutions, the analysis concentrates on potential for natural ventilation. In this context, the role of the underlying simulation engine in assessing the performance of the projected building is outlined.

Fig. 3 Example JAVA code defining EnergyPlus construction, glazing or control schedule input

```
EnergyPlus.defineConstruction("innerFacade construction",
                              "insulation 8cm",
                              "M15 200mm heavyweight concrete");

EnergyPlus.defineConstruction("floor construction",
                              "M15 200mm heavyweight concrete",
                              "insulation 8cm",
                              "wood 60mm");

EnergyPlus.setConstruction(glazing, "Dbl Clr 3mm/13mm Air");
EnergyPlus.setType(glazing, EnergyPlus.WINDOW);
EnergyPlus.setParent(glazing, zone.face(f));
```

3 Tools

The research makes use of an especially developed prototype framework building on *ANAR+*, an open source parametric scripting tool developed by the authors and used in architectural design studios [4]. This tool takes the form of a library addition to the popular coding framework *Processing.org* and its main characteristic is to concentrate on geometry definition through written code instructions (the *parametric script* of Fig. 2). The graphical user interface allows continuous alterations of parameters through *sliders*, whereas topological transformations must happen through code modifications. The geometric framework is extended by defining sets of higher level functions (e.g. an *API*, see Fig. 3) that provide interfaces to several expert-level BPS software, such as the whole building simulation EnergyPlus and its combination with the Building Control Virtual Test Bed (BCVTB) (Fig. 2, [10]).

This combination allows one to modify parametrically the geometry and run corresponding thermal, solar gains and airflow analysis providing the opportunity systematically to explore the influence of parameters on different design alternatives as often required in early design stage. This is done through automatic production of simulation input from the geometric description annotated with physical information in a fashion similar to Lagios et al. [5] and Toth et al. [9]. Given the source code availability, such an interface can be extended to access the full capability of the underlying BPS, addressing needs from non-experts up to advanced users. Further details may be found in [6].

Results may either be graphically represented or fed back to the parametric script for form modification according to performance — thereby allowing *performance-based* design — or for design-specific custom visualization building on the strength of the *Processing.org* project.

4 Case Study

As a design brief demonstrating the potential of the approach, the research considers an urban infill housing case study in a context typical to Berlin (Germany), with existing continuous square blocks and backyards, and facades possibly including oriels (Fig. 1). Although a common situation in the city, the context is abstract and does not correspond to a specific site.

The Potential of Scripting Interfaces 165

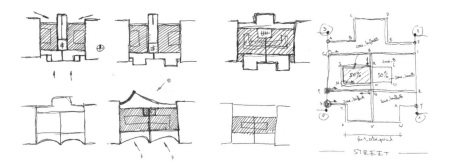

Fig. 4 Hand sketching study for alternative typology, *(left)* shading represents heated space, *(right)* rational sketch of chosen layout used for coding

Firstly, a reference model representing a standard design solution is defined and analyzed. Then, taking an example from the work of Lacaton & Vassal [7], an alternative typology is proposed that differentiates space on the outside edge of the building as non- or low-heated, acting as usable thermal isolation.

A parametric description is coded, expressing constraints such as global orientation or window facade proportion, with detail and material definitions required for analysis generated through code instructions. Thanks to the scripting approach, such detailing can easily evolve: using a simple method to initially populate the geometry, subsequent changes or refinements only require method adjustment.

Reference Typology

The reference case represents typical current building practice in Berlin. The continental climate requires significant insulation and double layer glazing, with a structure of reinforced concrete. The whole space is temperature-controlled (between 19° and 26°C). The parametric engine enables us to evaluate the impact of varying the facade glazing proportion on heating and cooling energy consumption (Figs. 5 and 6). An idealized HVAC plant model always able to meet heating or cooling demands is used, while in reality, space will either be warmer or cooler when installed power is insufficient.

To test the possibility of natural ventilation to achieve energy-free cooling, the ability of EnergyPlus to model airflow is available without additional input, thanks to automatic model translation. Non-experts can thus easily access and use this advanced technique while the coding interface allows further control of the simulation model. In such *airflow network* technique, complex flow patterns through openings are simplified to 1-dimensional two-way flows, representing a building as a network of flow paths. This strong simplification with regards to Computational Fluid Dynamics (CFD) computes larger buildings in reasonable time and, within known limits, provides meaningful results. Because of such models' complexity, non-expert

Fig. 5 Reference case: facade glazing proportion, variation from 20% to 50% *(left to right)*

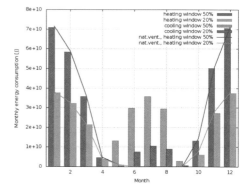

Fig. 6 Reference case: monthly energy needs, varying glazing proportion from 20% to 50%, natural ventilation cancel out cooling needs

users need to question results to obtain realistic consistency. To this end, the parametric property helps in conducting coarse *sensitivity analysis*. Using such models, it is clear from the case study yearly temperature frequencies (Fig 7), that natural ventilation achieves sufficient heat extraction even in the 50% glazing case.

The model for natural ventilation presented by EnergyPlus actually assumes automatic control, which has the advantage to always be present to take meaningful action. However, field studies have demonstrated that users accept warmer temperatures when they are able to themselves control windows for venting [2]. Using an additional interface to the BCVTB software [10], user control may be probabilistically modeled resulting in even cooler yearly temperature frequencies and therefore represents a reasonable design option (Fig 7). See [6] for additional details on the user model.

Alternative Typology

Inspired by the work of Lacaton & Vassal, an alternative typology is defined in which only a fraction of the surface is heated to standard comfort values, the remaining acting as buffer to protect from outside weather fluctuations. Usability of such buffer space is depending on weather, like a garden. The outer facade is fully glazed allowing rapid heating through solar radiation in winter, and can be fully

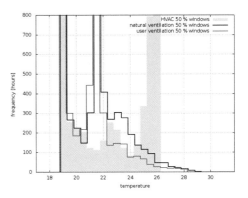

Fig. 7 Reference case: yearly temperature frequency for 50% glazing proportion with HVAC control, automated natural ventilation and user-controlled ventilation

opened to avoid overheating in the warm season. Here the typology layout needs to be further refined given the differing thermal conditions, which in terms of design process, directly sets usage in relation with performance.

After an initial in-depth sketch study (Fig. 4), a specific typology is chosen to be parameterized and analyzed through EnergyPlus. Initial results without airflow modeling (Fig. 9, left) convey the impression that buffer space would not be habitable due to overheating in summer; understandably, since the simulation model itself does not consider possible to open the outside glazing. However, modeling airflow makes buffer space follow outside temperature (Fig. 9, right), and reflects good comfort conditions (Fig. 11) with reduced energy consumption (Fig. 10).

A common trait in early design stage is to leave options open as long as they are not proven wrong. This often results in developing parallel design variants until one takes precedence. Would the operable glazing assessment of the alternative proposed not have been possible, chances are great for the option to be quickly ruled out, even more likely in a time-pressured practice environment.

Another common trait of early design is to open up the range of considered possibilities, including more experimental choices to push the project onwards: designers are taught to "think out of the box". However, the hidden normative inclination embedded in a simulation tool such as EnergyPlus whose limitations indirectly favour automatic control if not full HVAC are working against exploration of non-standard alternatives. The ability to systematically approximate thermal behaviour at early stages as well as to circumvent analysis tool limitations and hidden preferences, restrains designers from erroneously ruling out a valuable option. In this sense, tool flexibility for expansion and coupling is as important as ease-of-use to adequately support the design process.

Fig. 8 Alternative typology: inside glazing proportion, variation from 20% to 50% *(left to right)*

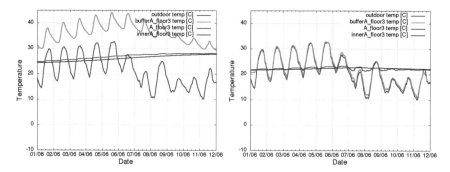

Fig. 9 Alternative typology: outside (red), buffer (green), middle (blue) and inner (fuchsia) space temperature time series during an early June heatwave, *(left)* without airflow, *(right)* with airflow modeling (3rd floor)

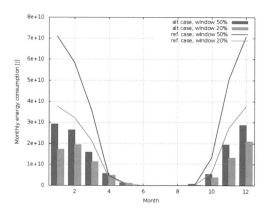

Fig. 10 Alternative typology: monthly energy needs, varying inside glazing proportion from 20% to 50% (3rd floor)

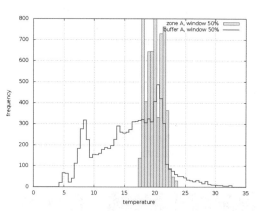

Fig. 11 Alternative typology: middle and buffer space yearly temperature frequencies for 50% glazing proportion (3rd floor)

5 Conclusion

Advantages of using a code interface both to describe form and run performance analysis have been presented. By comparing performance of standard and non-standard typologies for an urban infill context, the importance of the simulation tool capabilities has been unveiled. Correspondingly, the ability of the approach to circumvent inherent tool limitations by steering different levels of accuracies, as well as combining different analysis tools has been shown. Important implications for performance-based design strengthened by the designers' interest, especially at early design stage, in non-standard solutions have also been discussed.

Still in development, the research aims to enhance performance capabilities through additional tool modularity in order to more effectively support the design process. Such aptitude for modularity makes the parametric scripting approach a good candidate to empower designers to intertwine architectonic value with energy sustainability by conducting meaningful performance analysis.

Acknowledgements. Julien Nembrini and Guillaume Labelle are supported by the Swiss National Science Foundation grant numbers PA00P1_129120 and K-12K1-118078 respectively.

References

1. Bambardekar, S., Poerschke, U.: The architect as performer of energy simulation in the early design stage. In: Building Simulation 2009, 11th International IBPSA Conference, Glasgow, Scotland, pp. 1306–1313 (2009)
2. Haldi, F., Robinson, D.: On the unification of thermal perception and adaptive actions. Building and Environment 45, 2440–2457 (2010)
3. Hensen, J.L.M.: Towards more effective use of building performance simulation in design. In: Developments in Design & Decision Support Systems (2004)
4. LaBelle, G., Nembrini, J., Huang, J.: Programming framework for architectural design [anar+]. In: CAAD Futures 2009, Montreal, Canada (2009)
5. Lagios, K., Niemasz, J., Reinhart, C.F.: Animated building performance simulation (abps) - linking rhinoceros/grasshopper with radiance/daysim. In: SimBuild 2010, New York, USA (2010)
6. Nembrini, J., Labelle, G., Nytsch-Geusen, C.: Parametric scripting for early design building simulation. In: CISBAT 2011, EPFL, Switzerland (2011)
7. Ruby, A.: Lacaton & Vassal, Editions HYX, France (2009)
8. Schoen, D.: The Reflexive Practitioner: How Professionals Think in Action. Basic Books, New York (1983)
9. Toth, B., Salim, F., Drogemuller, R., Frazer, J., Burry, J.: Closing the loop of design and analysis: Parametric modelling tools for early decision support. In: Proc. Int. Conf. on CAAD Research in Asia (CAADRIA), pp. 525–534 (2011)
10. Wetter, M., Haves, P.: A modular building controls virtual test bed for the integration of heterogeneous systems. In: SimBuild, Berkeley, CA, USA (2008)

Building and Plant Simulation Strategies for the Design of Energy Efficient Districts

Christoph Nytsch-Geusen, Jörg Huber, and Manuel Ljubijankic

Abstract. This paper presents methods and strategies about how to model and simulate and optimize the energy demand and the energy supply of whole urban districts. For this purpose, different specialized modeling and simulation tools with different levels of detail, are combined to a higher strategy. To reach sufficient results for complex urban district models, this can be done for example with a variation in the level of detail of the physics, in the space and time resolution of the model or in the model accuracy. The application of these strategy is demonstrated by an use case of an energy infrastructure system for 2,000 new planned residential buildings in a 35 ha urban district, as a part of a New Town in northern Iran. Four different simulation based energy concepts were developed to obtain a highly primary energy demand and water demand reduction for the urban district. The best scenario demonstrates, that the use of energy efficient technologies together with an intensive usage of renewable energies (mainly solar energy) can reduce the primary energy demand by more than 65 percent and the water demand by more than 80 percent.

1 Introduction

After an intensive development and optimization of energy efficient building types during the last decades, the design of energy efficient districts or whole cities becomes more and more relevant. In particular, powerful analysis methods for the interrelations between the single buildings (e.g. their impact on the urban microclimate and on the mutual building shading) and simulation tools for the design of energy management systems on the district level (e.g. small district heating networks, combined heat and power production plants, renewable energy supply systems) are important core elements for this objective.

Christoph Nytsch-Geusen · Jörg Huber · Manuel Ljubijankic
Institute of Architecture and Urbanism, University of the Arts Berlin

For the design and the quantitative assessment of energy efficient districts it is necessary to model such building systems with all their relevant sub-components and control strategies in a well adapted physical and technical description. Unfortunately the widely-used very detailed building simulation tools are only suitable for the energetic design of single buildings or very small building groups. A detailed dynamic thermal building simulation for a urban district over long time periods (e.g. a year) with a huge amount of thermal consumers models, energy distribution models and energy plant models overstrains the present available simulation tools and computer hardware. For this reason, simulation models with a different level of detail are needed, to answer the bundle of questions, which occurs during the complex design and planning process of a energy efficient urban district:

First, for answering overall and strategic questions, e.g. the calculation of the yearly primary energy demand of a district, fast simulation system models are required, which are based on simplified sub-models for all the included building and building technologies.

Secondly, models for a more specific analysis of parts of the district or its energy infrastructure systems are needed, which have a higher level of detail in their space resolution and their physical description. Typical simulation scenarios are the optimization of critical technical components of the energy supply systems (e.g. the analysis of the thermal stratification of a centralized thermal storage of a district heating system) or the improvement of the energy efficiency of repeatedly used building types within the district (e.g. the thermal analysis of the base plates of the buildings).

Thirdly, sometimes two or more simulation tools have to be numerically coupled to a higher simulation approach. On the one hand, this could be combinations of complementary simulation tools, e.g. a thermal building simulation tool and a tool for HVAC (Heating Ventilation and Air Conditioning)-simulation and on the other hand combinations of detailed and simplified tools from the same domain (e.g. a 1D-simulation tool and a 3D-simulation tool for a sophisticated analysis of a thermo-hydraulic system).

2 Modeling and Simulation Levels

Four modeling levels of detail can be identified in the area of simulation of energy systems for buildings and building districts:

1. Models for building elements and energy plant components
2. Models for building zones and energy sub-systems
3. Models for buildings and energy supply systems
4. Models for building districts and integrated energy supply systems

The following examples for this four levels of detail shall demonstrate the range of typical questions and used simulation tools within an energy efficient design process for buildings and building districts.

2.1 Level 1: Building Element and Energy Plant Component

Figure 1 shows a typical scenario for the building element modeling. A 3D-thermal analysis of the base plate of a "German Passive House" together with its underlying concrete columns was done on the base of ANSYS CFD[1]. The simulation delivers 3D-temperature fields within the construction and 2D-temperature fields on its surfaces. Repeatedly interactive simulations with changes in the geometry and the material parameters let to a solution, where the heat loss through the base plate is only 8 percent higher in comparison to a ideal insulated plate without columns.

Figure 2 shows an example for an *Energy plant component modeling* by a transient 3D-analysis of the inner flow conditions of a thermal hot water storage of a solar thermal system, where ANSYS CFD was used for the calculation. The simulation was taken place with realistic boundary conditions of the surrounding energy plant, that means the hydraulic inlet of the storage was modeled with varying values of the mass flow and temperature [1]. In the picture, the situation during the early morning after the first switching on event of the load pump is illustrated. It is clearly visible how the heated fluid from the solar collector enters in the storage inlet with a high momentum. For example, such detailed hydraulic simulation can be used to find optimized shapes of the storage envelope, optimized geometries of the loading devices or optimized positions for temperature sensors within the fluid volume.

2.2 Level 2: Building Zones and Energy Sub-systems

Figure 3 illustrated a typical application for a detailed *Building zone modeling*. ANSYS CFD was used to simulate the air flow conditions for the heating case within a single thermal zone of a small residential building for two persons. The heat energy

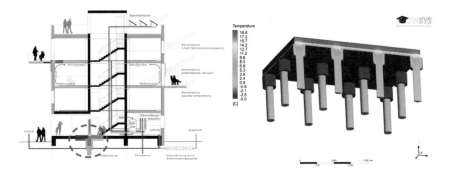

Fig. 1 Building element modeling: 3D thermal analysis of a base plate with columns. Position of the building element (left) and temperature fields of the vertical cut of the base plate (right).

---

[1] http://www.ansys.com/products/fluid-dynamics/cfd/

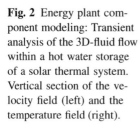
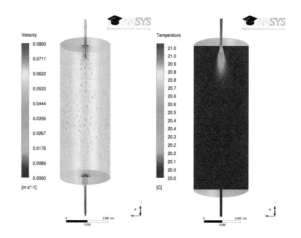

Fig. 2 Energy plant component modeling: Transient analysis of the 3D-fluid flow within a hot water storage of a solar thermal system. Vertical section of the velocity field (left) and the temperature field (right).

for the zone was supplied both with heated supply air and with a heat flux from a heating ceiling. The main question of the simulation was to determine the temperature and velocity distribution within the occupied area of the zone for an assessment of the thermal comfort. As the picture shows, the vertical temperature gradients and the velocities of the air guarantee good climate conditions.

Figure 4 shows an example for an *Energy sub-system modeling*. The object-oriented modeling description language Modelica[2] was used to model the thermal hydraulic flow conditions of a solar collector field as a part of a solar thermal system. Within this component-based modeling approach the energy plant is configured by the single technical components such as several solar thermal collectors, pipes,

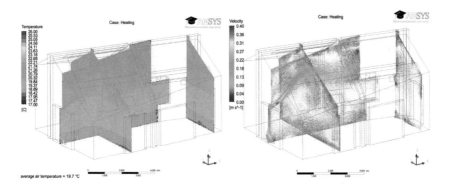

Fig. 3 Building zone modeling: simulated 3D-temperature- and velocity field of the room air of a heated zone

---

[2] http://www.modelica.org

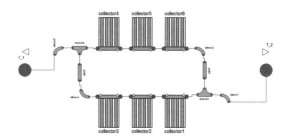

Fig. 4 Energy sub-system modeling: Component-based sub-model of a collector field as a part of a solar thermal system

branches and elbows. With such kind of sub-system model, the impact of a hydraulic failure within one of the collectors on the energy gain of the total collector field can be studied [2].

2.3 Level 3: Building and Energy Supply System

Figure 5 (left) shows an example for *Building modeling*, where a set of row houses were analyzed in terms of their heating and cooling demand within the urban planning project "Young Cities" [3]. The building shape, the building construction and the thermal zoning was modeled with Autodesk Ecotect[3]. To obtain a high accuracy in the simulation results, the model data was exported to the building simulation tool EnergyPlus[4]. This multi-zone building models are typically used to calculate timelines of the heating and cooling load, for each single zone and also for the total building taking into account all interactions between the thermal zones.

Figure 5 (right) illustrates an example for *Energy supply system modeling*. The system model of a solar thermal plant was modeled again with Modelica, where

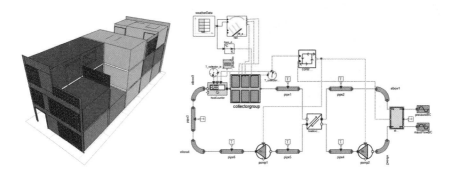

Fig. 5 Building modeling (Thermal multi-zone building model) and Energy supply system modeling (System model of a solar thermal plant)

---

[3] http://www.autodesk.com
[4] http://apps1.eere.energy.gov/buildings/energyplus/

the complex structured energy sub-system model of the collector field from Figure 4 is reused as a component model [2]. On the system level, questions regarding to the interrelations between the controller devices, the controlled systems under the boundary conditions of the local climate can be studied. Another typical question for this model type is the determination of the most important system parameters (e.g. the storage volume, the number and orientation of the collectors or the variation of the collector type) with the aim of a high energy efficiency on the system level.

2.4 Level 4: Building District and Integrated Energy Supply System

Figure 6 (left) shows a thermal building model for a sub-neighbourhood, again from the Young Cities project as a typical application for *Building district modeling*. The 20 row buildings, grouped in four blocks, are modeled and simulated with the "tool-tandem" Autodesk ECOTECT/EnergyPlus, where each building is modeled simplified as one thermal zone. This model type is suited to analyze the mutual shadowing of the single buildings and serves also as a calculation base for integrated energy supply concepts, because it delivers time-dependent data for the heating and cooling load of each building.

Figure 6 (right) illustrates a complex example for an Integrated energy supply system modeling. The district heating network of the small German town Rheinsberg was modeled with the simulation system SMILE, where the 500 thermal consumers - the heated buildings - were represented as strong simplified thermal building models [4]. Based on this model, critical locations of the thermal energy supply during cold winter days can be identified within the network (look at the locations of the blue visualized return pipes of the network).

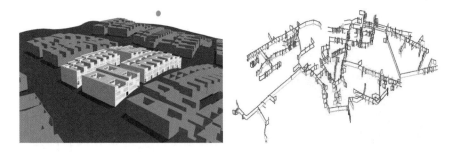

Fig. 6 Building district modeling (Thermal model of twenty buildings of a sub-neighbourhood) and Integrated energy supply system modeling (Model of a district heating net of a small German City, which supplies 500 thermal consumers)

3 Coupling of Simulation Models and Tools

Normally, an integrated simulation approach with one common numerical solver for the complex system model provides the best performance for the simulation experiment. Often, different simulation tools with a complementary content, but a different software architecture, have to be work in cooperation. In these cases software frameworks for numerical coupling can integrate models from one or more software platforms to a system model on a higher level.

Figure 7 shows the application of two of such frameworks in the field of energetic building simulation. In the approach on the left side, the Middleware TISC [6, ?] was used to realize a numerical coupling between a Modelica system model of a solar thermal plant with the CFD-component model (the model from Figure 2) of the hot water storage [1].

The approach on the right side uses the BCVTB (Building Control Virtual Test Bed), which is an open source co-simulation software environment [7]. Here, a heating system, based on Modelica/Dymola and a multi-zone thermal building model with 11 zones (the model from Figure 5 left), based on Energy Plus are combined to entire building and plant simulation model.

4 Use Case Project Young Cities

The following use case concerns about the simulation methodology within the Young Cities Research Project. In this project, a new district for 8,000 inhabitants in Hashtgerd (Iran) is designed and planned. The entire energy efficiency and particular the efficiency for the energy infrastructure and building technology has a high priority. Four different efficient energy supply systems - central, semi-central and de-central variants - were designed and compared under each other regarding their primary energy and water demand. Two of these systems are shown in Figure 9 and 10. A reference system (a conventional Iranian energy supply system) was simulated for the assessment of the new designed systems.

By the use of the bandwidth of the described simulation models und methods in last two sections, the specific primary energy demand and water demand for heating

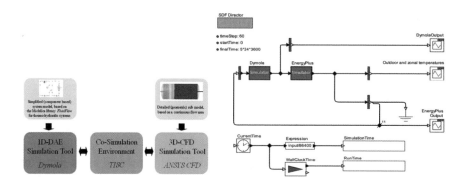

Fig. 7 Two approaches for the numerical coupling of simulation tools. Left: 1D-/3D-coupling of the tools Modelica/Dymola with ANSYS CFD, based on TISC. Right: 1D-/1D-coupling of the tools Modelica/Dymola with EnergyPlus, based on BCVTB.

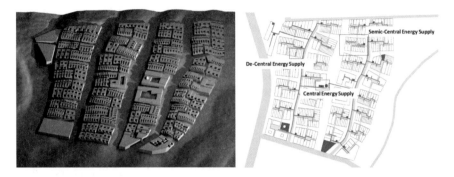

Fig. 8 City model of the new planned 35ha district for 2,000 living units (left) in Hashtgerd New Town (Iran). Within the project, variants of de-central, semi-central and central energy supply systems are being developed for the heating, warm water and cooling demand of the buildings (right)

Fig. 9 Improved "Conventional Iranian energy supply system" on de-centralized level (single buildings), based on cold water production with photovoltaic modules and compression chillers and the use of solar thermal energy for warm water production and heating assistance

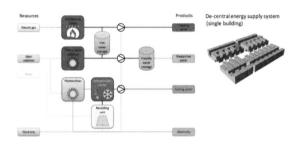

and cooling for all scenarios were calculated. The best scenario, the photovoltaic based energy system from Figure 10 can reduce the primary energy demand by more than 65 percent and the water demand by more than 80 percent.

5 Conclusion and Outlook

For a simulation based design of energy efficient districts, first of all a lot of information has to be considered. For this reason, GIS-tools and in addition digital city models can be used to collect, structure and store these data as an input for the following energetic building and district simulation analysis. Today, a bundle of simulation models, tools and methods, differentiated in their level of detail are available for this purpose. In many cases, it makes sense to integrate two or more simulation tools by a numerical coupling on a higher level in order to reflect all

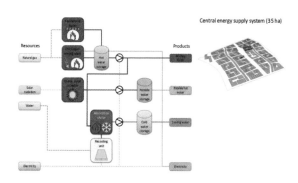

Fig. 10 Energy supply system for the total 35 ha area, based on a central energy production with co-generation (electricity and thermal energy from natural gas) and a de-centralized production of cooling energy with absorption chillers (heat from solar radiation and backup energy from the district heating network)

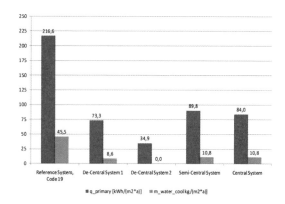

Fig. 11 rea specific primary energy demand and water demand for the analyzed variations of energy supply systems for the 35 ha district in Hashtgerd New Town (Iran)

important interrelations between the buildings, the energy infrastructure systems and the local climate conditions of the district.

In the next years, the development of simulations models for energy efficient districts will be deepened, especially simulation tools with a balanced modeling depth in the entire district model have to be developed. The CitySim project from the research group at EPFL represents an ambitious approach in this direction [5].

References

1. Ljubijankic, M., Nytsch-Geusen, C., Rdler, J., Lffler, M.: Numerical coupling of Modelica and CFD for building energy supply systems. In: Proceedings of 8th International Modelica Conference, Dresden, March 20-22 (2011)
2. Nytsch-Geusen, C., Ljubijankic, M., Unger, S.: Modelling of Complex Thermal Energy Supply Systems Based on the Modelica-Library Fluid Flow. In: Proceedings of 7th International Modelica Conference, Como, September 20-22 (2009)

3. Ederer, K., Huber, J., Nytsch-Geusen, C., Seelig, S., Unger, S., Wehage, P.: Konzeption und Planung solarunterstützter Energieversorgungssysteme für New Towns im Iran. In: Tagungsband: 20. Symposium Thermische Solarenergie in Staffelstein, OTTI-Technologiekolleg, Regensburg (2010)
4. Klein-Robbenhaar, C., Kcher, R., Nytsch, C., Waldhoff, C.: Numerische Simulation von Wrmenetzen. Simulation des Betriebsverhaltens von Wrmenetzen mit Integration der versorgten Gebude. In: Brennstoff-Wrme-Kraft .VDI-Verlag, Dsseldorf (September 2000)
5. Robinson, D., Haldi, F., Kmpf, J., Leroux, P., Perez, D., Rasheed, A., Wilke, U.: CitySim: Comprehensive micro-simulation of resource flows for sustainable urban planning. In: Proceedings of Eleventh International IBPSA Conference Glasgow, Scotland, July 27-30 (2009)
6. Kossel, R., Correia, C., Lffler, M., Bodmann, M., Tegethoff, W.: Verteilte Systemsimulation mit TISC. In: ASIM-Workshop 2009 in Dresden mit integrierter DASS (2009)
7. Wetter, M., Haves, P.: A modular Building Controls Virtual Test Bed for the integration of heterogeneous systems. In: Prodeedings of Third National Conference of IBPSA-USA, Berkeley, California, July 30 -August 1 (2008)

New Design and Fabrication Methods for Freeform Stone Vaults Based on Ruled Surfaces

Matthias Rippmann and Philippe Block

1 Introduction

Thin concrete and steel grid shells show elegantly how shell design is used for contemporary freeform architecture. Their natural beauty is coupled to an inherent efficiency due to minimal bending, result from their good structural form. Thanks to digital form finding tools, streamlined planning processes and automated fabrication, the technical and economic difficulties to design and build those structures, especially grid shells, decreased significantly [1].

In contrast, the use of stone as a structural material for elaborate and exciting freeform architecture did not develop in a similar manner. However, even today, Gothic cathedrals show the natural aesthetics of stone structures in an impressive way, combining structural and ornamental building parts to realize complex building forms [2]. The industrialized methods of processing stone, needed for freeform vaults, are less established and less flexible than comparable procedures for steel and concrete, for which a continuous and great improvement took place over the last century [3]. The new approaches shown in this paper aims to pave the way for the use of stone for freeform vault structures in the near future.

In this context, ruled surfaces are of particular interest due to the geometric configuration of discrete stone blocks (voussoirs) of freeform vaults, which originates both from the statics of vaults (Section 2.1), and from stone cutting based on wire-cutting technology (Section 2.2).

The use of ruled-surface design has been of special interest for freeform architecture, which comprises parts or patches of ruled surfaces, for example of hyperbolic paraboloids (hypar), rotational hyperboloids or helicoids [4]. Firstly, ruled surface based design is used to explicitly describe those forms, which dates back to the work of Antoni Gaudí, who was a pioneer in using ruled surfaces to define the complex geometry of his designs (fig. 1.1 a) [5].

Matthias Rippmann · Philippe Block
Institute of Technology in Architecture, ETH Zurich, Switzerland

A second, well known use of ruled surfaces in the building process is for rationalizing the construction work for a building. Felix Candela for example took advantage of the fact that through every point on a ruled surface runs at least one straight line that lies on this surface. Therefore, he constrained his thin, concrete shells to combinations of parts of hypars which then could be built using formwork out of linear elements (fig. 1.1 b, right). Moreover, doubly ruled surfaces can be used to apply planar quadrilateral panels on freeform architecture (fig. 1.1 b, left) [4].

Thirdly, fabrication techniques based on ruled surfaces, such as diamond-wire for stone or hot-wire cutting for foam, typically process these material faster, with less kerf and material waste, than traditional subtractive manufacturing methods such as milling and solid blade machining [6]. For example, for the glass fiber reinforced concrete panels of the facade of the Cagliari Contemporary Arts Center by Zaha Hadid Architects hot-wire cutting was used to produce the EPS foam molds for the freeform panels (fig. 1.1 c) [7].

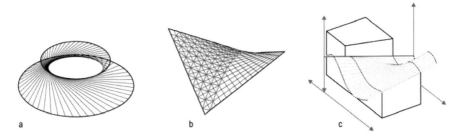

Fig. 1.1 (a) The definition of a hyperboloid by a moving line along two circles (b) A hyperbolic paraboloid consists of two families of intersecting lines and can be populated with planar quad meshes (c) A tensioned wire cuts through material, tracing a ruled surface

For the design and realization of freeform stone vaults we considered those strategies to consequently identify efficient ways of feasible construction. A key aspect was to define and develop a suitable and coordinated design and fabrication setup for the production of hundreds of individual voussoirs, which need to be processed for a single vault design. Due to the three-dimensional shape of the voussoirs and the geometrically complex fabrication constraints, the challenge is to coordinate the design of the individual voussoirs in combination with the technical machine setup. The challenge is to find the right balance of the integrated design and fabrication method to achieve a feasible configuration to produce freeform stone vaults efficiently. This potential of a well-coordinated digital design and fabrication setup for cutting stone to realize freeform stone vaults has not been tapped yet. Additionally, the software to design, simulate and process the part geometry is typically used one after the other, missing an integrated concept to cope with the above mentioned challenges.

In this paper, we show that new approaches of digital design and fabrication methods unveil great potential for building processes in the field of freeform stone vaults. Considering the materialization process of freeform vaults the research leads to a novel, streamlined approach, demonstrating the smooth integration of structural needs, fabrication constraints and economic feasibility into the design and production process.

2 Geometrical Considerations

In this section, the geometrical and technical information about discrete freeform vaults and state-of-the-art wire-cutting machine setups will be described. Also, the relevance of ruled surfaces for freeform stone vault design and fabrication will be described.

2.1 Freeform Stone Vaults

The design and realization of freeform vaults comprises four parts: form finding [8], tessellation [9], voussoir generation, also known as stereotomy [10], and vault assembly.

This Subsection focuses on the generation of the voussoirs, which is based on the tessellation and the surface geometry of the funicular form. Ideally, to prevent sliding failure and guarantee structural stability, the main load bearing faces of neighboring voussoirs need to be perpendicular to the force flow. Therefore, the interface surfaces should be normal to the thrust surface at any point. This can be best described by examining a single voussoir based on a hexagonal tessellation pattern applied on an arbitrary thrust (i.e compression-only) surface as shown in fig. 2.1.

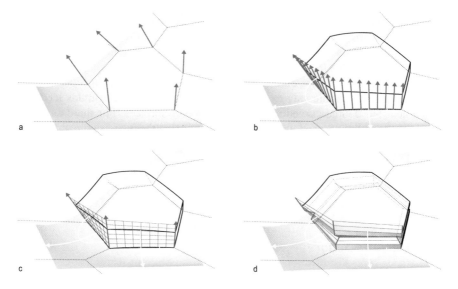

Fig. 2.1 (a) Surface normals define the vertical edges of voussoirs (b) A family of surface normals defines the load bearing surface (c) Lofting between two edge lines results in a doubly ruled surface (d) The horizontal orientation of the generators increases the geometric flexibility

The tessellation pattern is represented by a set of points on the surface and a network of connecting lines. Considering the perpendicularity to the thrust surface, the orientation of the vertical edges of one or more voussoirs are defined by the surface normal at each of those points (fig. 2.1 a). Consequently, a family of surface normals between those vertical edges defines a ruled surface. This can be achieved by considering all surface normals on a geodesic curve between two corresponding points on the surface (fig. 2.1 b). Considering the fabrication constraints however, approximation through lofting between two vertical edge lines can be advantageous (fig. 2.1 c). First, the resulting doubly ruled surface can be traced by a wire in two directions, hence allowing more flexibility in the fabrication process. Second, the horizontal orientation of the generators allows for integrated interlocking notches, used for later reference during the assembly of the individual voussoirs (fig. 2.1 d). Due to the relatively small surface curvature between two points, the deviation between both alternative surfaces (fig. 2.1 b, c) can be neglected.

The previous approaches for generating the interface surfaces of voussoirs can typically not be applied in the same manner to the upper (extrados) and lower (intrados) surfaces of the voussoirs, due to their freeform, double-curved shape which can normally not be replaced by ruled surfaces, and hence, cannot be fabricated using a wire saw. A possible solution is to use an additional fabrication technology such as 5-axes milling or circular saw blade stone cutting. This will guarantee a smooth continuous extrados and intrados of the freeform stone vault (fig. 2.2 a). In contrast to the interface surfaces, the precision of the intrados and extrados is structurally of less importance, as long as the required vault thickness is guaranteed and thus can be approximated by ruled surfaces. This can be done by a global approximation of the intrados and extrados by strips of ruled surfaces [7]. This global approach would need to be highly coordinated with the tessellation in order to prevent generators of one ruled surface from intersecting with other parts of the voussoir geometry, especially for surfaces with negative curvature, as that would make the use of wire-cutting technology impossible.

An alternative, simple way to approximate the intrados and extrados locally is to segment the double-curved surfaces by two part surfaces. For the case of the hexagonal tessellation used as example in this paper, these are two quadrilateral surfaces (fig. 3.2 c). In the process shown in this paper, they will be automatically aligned to best fit the original geometry, regarding the distances between the midpoints to the corresponding closest points on the original intrados and extrados. If needed, this kinked geometry can be finished smoothly after installation by semi-manual chiseling techniques.

2.2 Machine Setup

In order to provide the appropriate fabrication setup to process the geometry described above, different profiling machines have been examined in detail. They are equipped with a diamond wire to cut the stone within a variety of possible machine configurations. They range from single wire saws for dressing blocks and cutting slabs to computer-controlled systems for two- and three-dimensional profiling [11]. Those machines consist of standard elements, but customized configurations in terms of size and geometric flexibility are quite common. The most relevant machine configurations for stone processing based on wire-cutting technology can be categorized by the configuration and number of independent axes (fig. 2.2).

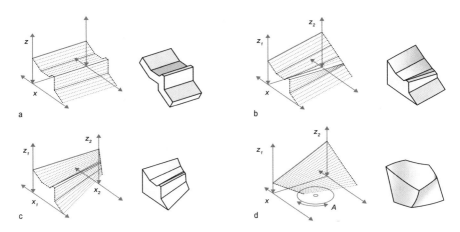

Fig. 2.2 (a) Profile cuts with a 2-axes machine (b) Conoid cuts with a 3-axes machine (c) Conical directed cuts with a 4-axes machine (d) Conical multidirectional cuts with a 4-axes machine

Simple stone profiling can be processed on a 2-axes machine setup (fig. 2.2 a), which enables planar cuts and is mostly used to cut rough slabs along the X and Z axes. More complex cuts are possible by using two independent vertical axes (Z_1, Z_2) (fig. 2.2 b), e.g. to produce elements based on conoids where all rulings are parallel in planar projection. In contrast to the 2-axes setup, the wire length changes caused by asynchronous movement along the two vertical axes, raises the technical requirements for the wire guide and tensioning system.

For conical cuts a forth axis is needed (fig. 2.2 c). The combination of two individual horizontal axes (X_1, X_2) and two individual vertical axes (Z_1, Z_2) offers great geometrical flexibility, especially for linear elements. The increasing problem of the length differences for diagonal wire positions might be one reason why no wire saws based on this configuration could be found by the first author. In contrast, the same configuration is often found for CNC hot-wire cutters.

Volumetric parts with multidirectional orientation of ruled surfaces can be processed by the machine arrangement with only one horizontal axis (X) and two individual axes in the vertical direction (Z_1, Z_2), the forth axis (A) in this case defines the rotation of a turntable (fig. 2.2 d). Even more geometrical flexibility is possible using wire-cutting based on robotic arm fabrication [12,13], but due to the limits of weight and picking mechanisms, the potential for this technique lies in hot-wire foam cutting, rather than in diamond-wire stone cutting and was therefore not being examined further.

3 Experimental Design and Fabrication Setup

Considering the geometrical challenges of freeform stereotomy, the research shown in the previous chapter identifies the 4-axes machine configuration with one rotational axis (fig. 2.2 d) has the most appropriate and efficient setup. The configuration allows the fabrication of complex volumes represented by patches of ruled surfaces. However, the relatively high investment needed to experiment with real stone cutting, led to the development of a customized CNC hot-wire cutter. It allows the exploration of automated foam cutting as a simulation of CNC stone processing. Hence, the ongoing research focuses on the geometrical constraints of those processes, rather than on the technical challenges of stone cutting related to the large weight of the material or the complexity of the abrasive cutting process.

Nevertheless, the possible size and shape of a voussoir in relation to the machine clearance, angle restrictions, and the stone volume, is very difficult to grasp. Therefore, a plugin for Rhinoceros 5.0 was developed to control conflicting parameters and visualize the result, respectively simulate the production process. This information was then evaluated and used to develop the machine setup and the prototypical voussoirs shown as a result of this research.

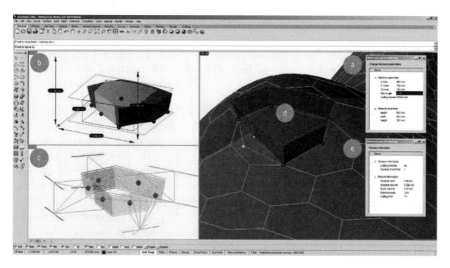

Fig. 3.1 While manipulating design, fabrication and material parameters, the production process is analyzed, visualized and simulated using a specially developed plugin for Rhinoceros 5.0

The screenshot shown in Figure 3.1 shows the CAD environment with the customized panel to control the parameters (fig. 3.1 a) and the visualization of the results (fig. 3.1 b,c). Machine parameters, such as individual axis length, maximum wire angle and the cutting speed, are defined numerically in a specific panel (fig. 3.1 a) but can also be manipulated by moving the endpoints of the axes in 3d (fig. 3.1 b). The material dimensions can be changed in a similar way.

Besides the virtual machine configuration and the processed voussoir (fig. 3.1 b,c), the tessellation and voussoir geometry on the thrust surface (fig. 3.1 d) show the individual voussoir in the context of the overall vault geometry. This is used for real-time manipulation by changing the tessellation topology interactively on the thrust surface. This will affect the shape and dimension of the voussoir, which is automatically aligned and positioned within the virtual building chamber of the machine (fig. 3.1 b). This real-time positioning is based on the minimum bounding box of the voussoir and the steepest generators of the corresponding ruled surfaces. The visualization of the aligned voussoir (fig. 3.1 b), tool path information (fig. 3.1 c) and additional coloring, indicates possible fabrication conflicts for the chosen configuration. Moreover, the numerical output shows the number of conflicts, the amount of waste material and the approximated cutting time (fig. 3.1 e).

The developed program was used to verify and improve the design and fabrication setup described in this paper. In particular, the geometry of a previously developed freeform thin-tile vault [14] was used to test and verify the described methods, from the initial design to the final production, for the same geometry built from discrete stone blocks (fig. 3.2).

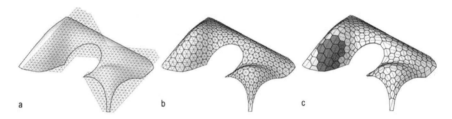

Fig. 3.2 (a) The NURBS surface and a planar network of triangular elements (b) The network gets evenly distributed on the surface using customized relaxation methods [9] (c) Depending on the curvature of the surface, the hexagonal and quadrilateral voussoirs are being generated

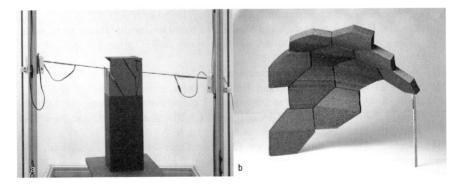

Fig. 3.3 (a) Shows the cutting process of a single voussoir. The Styrofoam block is mounted on the turntable, while the hot wire is travelling through the material. (b) This voussoir sample is one single piece of a patch of highly individual part geometries, also shown in fig. 3.2 c

4 Conclusion

This paper has described a powerful approach and technique to materialize stone vaults, considering today's high demands on feasibility and effectiveness. A strategy for generating individual and geometrically complex voussoirs has been developed, taking into account architectural, structural and fabrication requirements. An appropriate fabrication setup in combination with the design method was developed and further analyzed and improved using a custom-made plugin for the visualization of the complex interdependencies of design, fabrication and material parameters. Thanks to this program, it was possible to demonstrate that different freeform vault designs can be generated and produced based on ruled surface geometry. Hence, allowing the use of hot-wire cutting, which persuasively illustrates the possibilities for real stone cutting. Moreover, the real-time feedback about relevant fabrication data helped to control the feasibility of the process prior production and finally facilitated the technical realization.

Future research will focus on an automated feedback system for the voussoir generation, which not only visualizes the limitations of the fabrication, but automatically finds the best possible configuration considering specific criteria. This process could be done iteratively by using genetic algorithms or nonlinear optimization methods. Therefore, the speed and real-time performance of the program needs to be improved. Concerning the realization of real stone vaults, possible industry partners have already been contacted to work on the technical implementation of the stone-cutting process.

References

[1] Glymph, J., et al.: A parametric strategy for free-form glass structures using quadrilateral planar facets. Automation in Construction 13(2), 187–202 (2004)
[2] Evans, R.: The Projective Cast. MIT Press, Cambridge (1995)
[3] Schodek, D., Bechthold, M., et al.: Digital Design and Manufacturing: CAD/CAM Ap-plications in Architecture and Design. Wiley, Hoboken (2005)
[4] Lordick, D.: Intuitive Design and Meshing of Non-Developable Ruled Surfaces. In: Proceedings of the Design Modelling Symposium 2009, October 5-7, pp. 248–261. University of the Arts, Berlin (2009)
[5] Burry, M., et al.: Gaudí unseen: completing the Sagrada Família. Jovis, Berlin (2007)
[6] Asche, J.: Tiefschleifen von Granit. University of Hannover, Dissertation (2000)
[7] Flöry, S., Pottmann, H.: Ruled Surfaces for Rationalization and Design in Architecture. In: Proceedings of ACADIA 2010, pp. 103–109. The Cooper Union, New York (2010)
[8] Block, P.: Thrust Network Analysis: Exploring Three-dimensional Equilibrium. Dissertation, Massachusetts Institute of Technology, Cambridge (2009)
[9] Lachauer, L., Rippmann, M., Block, P.: Form Finding to Fabrication: A digital design process for masonry vaults. In: Proceedings of the International Association for Shell and Spatial Structures (IASS) Symposium 2010, Shanghai (2010)

[10] Rippmann, M., Block, P.: Digital Stereotomy: Voussoir geometry for freeform masonry-like vaults informed by structural and fabrication constraints. In: Proceedings of the International Association for Shell and Spatial Structures (IASS) Symposium 2011, London (2011)

[11] Pellegrini: Robot Wire Stone Profiling. Catalogue Profiling Machines (2011), http://www.pellegrini.net/cataloghi/PELLEGRINI_RobotWire.pdf (accessed June 10, 2011)

[12] Gramazio, F., Kohler, M.: Designers Saturday. Design exhibition, Langentahl, Digital Fabrication in Architecture, ETH, Zurich (2010), http://www.dfab.arch.ethz.ch/web/d/forschung/191.html (accessed June 10, 2011)

[13] Meier, M.: Robotic Fabrication of Parametric Chairs. Taubman College of Architecture. University of Michigan, Ann Arbor (2011), http://mkmra2.blogspot.com/2011/04/robotic-fabrication-of-parametric.html

[14] Davis, L., Rippmann, M., Pawlowfsky, T., Block, P.: Efficient and Expressive Thintile Vaulting using Cardboard Formwork. In: Proceedings of the International Association for Shell and Spatial Structures (IASS) Symposium 2011, London (2011)

Design and Optimization of Orthogonally Intersecting Planar Surfaces

Yuliy Schwartzburg and Mark Pauly

1 Introduction

We present a method for the design of 3D constructions from planar pieces that can be cut easily and cheaply with laser cutters and similar Computer Numerical Control (CNC) machines. By cutting tight slits in intersecting pieces, they can be slid into each other forming stable configurations without any gluing or additional connectors. These constructions enable quick prototyping and easy exploration of shapes, and are particularly useful for education. We propose a constraint-based optimization method and computational design framework to facilitate such structures.

Planar surfaces can be connected, without any other materials or bindings, by cutting slits of the width of the material in the direction of the surface normal. These constructions are often found in cardboard and wooden 3-D puzzles (see Fig. 2). As laser cutters do not permit cuts that are not orthogonal to the surface, the slits would need to be larger to accommodate intersecting pieces at non-right angles (see Fig. 3). This would lead to unstable configurations and the need for connecting structures. Our goal is to avoid this complication and retain the simplicity of orthogonally intersecting pieces.

As shown in Fig. 3, when considering a pair of intersecting orthogonal pieces, each piece has one degree of freedom in rotation. This introduces a constrained design space for composing more complex shapes from multiple interconnected pieces. The common method for generating such objects is using parallel or semantic cross-sections (as in Fig. 2), but this severely limits the design space. Slightly more complex methods grow the structure generatively, starting with a skeleton and ensuring orthogonality step by step, in either a manual or procedural fashion. However, these approaches quickly become infeasible due to combinatorial complexity

Yuliy Schwartzburg · Mark Pauly
Computer Graphics and Geometry Laboratory (LGG) EPFL Lausanne, Switzerland

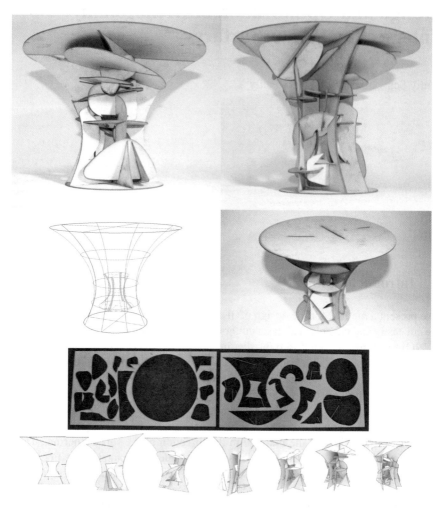

Fig. 1 A physical table built with orthogonally intersecting planar pieces using our design method. Also pictured are the hull wireframe, the negatives of the corresponding pieces, and the building process.

inherent in the global coupling of orthogonality constraints. The design space is difficult to predict and one design step can have unforeseeable implications for the end product (as seen in Fig. 5).

2 Related Work

The utilization of computational techniques during the design process has been studied extensively, and in particular, in the context of performative and construction-aware design. Axel Killians doctoral thesis focuses on the interrelation between

Design and Optimization of Orthogonally Intersecting Planar Surfaces

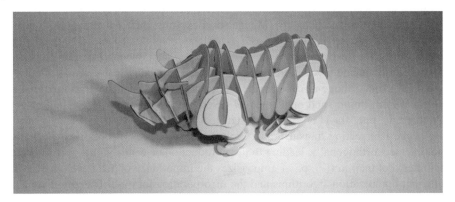

Fig. 2 An existing commercial cardboard artwork. Note that this was designed according to the skeleton of the rhinoceros and its cross-sections do not exhibit much variation in orientation.

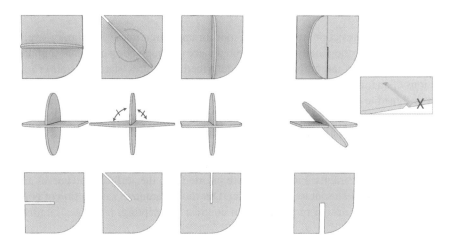

Fig. 3 Orthogonal intersection of two planar pieces allows rotation of the pieces relative to each other along one axis without violating orthogonality constraints. The first row shows assembled pieces while the second row shows a side view, in which the 90-degree intersection is apparent. The bottom row shows the cutting curve of the bottom piece. Fixing the bottom piece, the green arrow represents valid rotations of the second piece, as can be seen in the first three columns, which still form intersections of 90 degrees. All other rotations (red arrow) are not allowed. The fourth column represents one such invalid orientation. To produce tight, stable slits, the result should look like the inset piece. However, this is not easily accomplished with a laser cutter. Therefore, much wider, unstable slits are introduced.

design and constraints [3]. Oxman explores the link between performative design and computational geometry [6]. Mangelsdorf explores different strategies to deal with complex geometries and praises a hybrid approach that enables a high degree of freedom in the development of the form, but integrates concepts based on physics, form description and fabrication. Particularly, the Médiacité Liège consists of intersecting (but not orthogonal) planar rib sections that were designed with a mixture of physical form-finding and mathematical descriptions ([4], pp. 41-45). The Sphere Project ([1], pp. 103-111)) explores intersections of planar surfaces. They generate planar intersecting rings using an evolutionary process, however, their rings do not intersect orthogonally, leading to the introduction of welds, joints, and discontinuities. Pottman in ([7], p. 74)emphasizes the importance of"the development of efficient optimization algorithms and the incorporation into user-friendly rationalization software tools." While our design space is too simplistic to fully capture the complexity of large-scale architectural projects, it nevertheless allows study of how computation and optimization can be leveraged to enable an effective design process.

3 Design Process

We present a construction-aware design tool for enabling effective design of structures containing orthogonally interconnected pieces through computational support. The tool facilitates the building of the described structures starting from cross-sections of an initial manifold mesh or boundary representation that defines the hull. It is designed to be time-saving and intuitive while giving freedom to the architect to prescribe or break necessary constraints. We use an iterative design process in which the designer places planar surfaces within the desired volume and a feasible solution is calculated using an optimization approach based on constraint satisfaction. As the design process continues, the planes are updated based on the coupling defined by the constraints. Finally, slits are inserted into the surfaces and curves are produced, ready to be sent to a laser cutter. Our tool guarantees satisfaction of constraints during the design process, and in contrast to scripting, facilitates intuitive design handles.

3.1 Optimization Constraints

For details on the constraint satisfaction solver and its implementation we refer to [2]. In this paper, we concentrate on the design process of utilizing such a solver. It is designed to find a feasible solution satisfying all necessary (hard) constraints (in our case, orthogonality) while remaining close to the original design. Additional design criteria can be specified in the form of soft constraints. These constraints will be satisfied as well as possible within the design space defined by the hard constraints. Additional constraints necessary for a specific model can be programmed and inserted into the solver (stability, tension, fixed boundaries, daylighting, heat distribution, etc.).

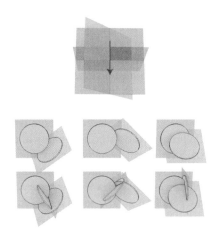

Fig. 4 If two planes are fixed, there is only one allowed orientation for a third plane connecting the two. The purple and green planes form an intersection line (red) that uniquely determines the normal direction of the plane connecting the two (yellow). Three examples are illustrated, before and after adding the third plane.

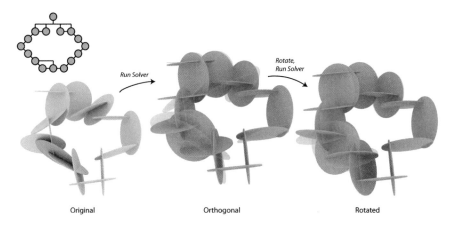

Fig. 5 A cyclic arrangement of pieces. The graph representing the connections is shown at top-left. Beginning with a non-orthogonal construction at left, we run the optimization solver. Then, as a user edit operation the yellow piece is rotated and the optimization runs again. The transparencies in green illustrate the previous locations of the pieces for comparison. Note that rotating the yellow piece produces changes in many neighboring pieces, and slight changes in far pieces.

3.2 Orthogonality

The basic hard constraint of our system is that two intersecting pieces must meet at a right angle. If we look at a system of two pieces (A and B), if we fix piece A we leave one degree of freedom for piece B, rotation around the normal of A. Consider a system of three pieces (A, B, and C) with B intersecting A and C, but A and C independent. If we fix A and C and A is not parallel to C, then B has a fixed orientation (see Fig. 4) If A is parallel to C, then this can be treated as the first case.

To represent the global configuration, we can consider a bidirectional graph of connections $G = (V, E)$, where a vertex V represents a planar pieces and an edge E represents an intersection between two pieces. While it is not difficult to ensure orthogonality at the start of the design process in an iterative approach, as cycles are introduced into G, a new intersection or a change in orientation can propagate through the graph, requiring modification of a number of pieces. However, by using optimization techniques, we can automatically calculate the minimal necessary modification of the placed pieces in order to satisfy the constraints.

In order to satisfy orthogonality in the optimization, we need to rotate each non-orthogonal intersecting pair as explained in Figure 6. The solver iterates, each time satisfying constraints and then merging the results, until a consistent state is reached. It is often not possible to satisfy every intersection (or the results become uninterestingly parallel); an unwanted intersection can then be eliminated by cutting a piece at a set distance from the intersection.

We use an iterative approach of optimizing after placing each plane. The optimization can be run once at the end of the design process as a post-rationalization as well (see Fig. 5). However, there is less chance of deviating far from the intended design when optimizing iteratively.

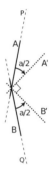

Fig. 6 Consider two pieces, A and B, defined on planes P and Q respectively. We take the intersection of P and Q as the rotation axis. Then, we rotate the pieces around their respective centroids (in opposite directions) by $a/2$, where a is 90 degrees subtracted by the angle between P and Q

3.3 Position

By itself, the orthogonality constraint would rotate each piece such that globally, the movement is minimal. However, a necessary constraint in many designs is to lock down certain integral pieces such that the optimization does not modify them unnecessarily. A position constraint locks a piece to its original position such that the algorithm will strive to solve the optimization without moving that piece. This can be useful to define load-bearing pieces or pieces designed to break other constraints. Silhouette.

As an example of a performative constraint, we consider pieces forming a set silhouette (see Fig. 8) when viewed or lit from a certain direction [5]. This soft constraint can be used for an application such as daylighting or can enable the designer to intuitively fill a section of a certain volume for semantic purposes. The constraint can also be prescribed to necessitate a certain shape or "skeleton" in order to maintain structural integrity. Consider two intersecting pieces A and B. The silhouette constraint is satisfied locally within the solver by translating A and B in opposite directions such that they maximally fill the space defined by the projected silhouette.

3.4 Movement

A piece can additionally be restrained to a certain volumetric region, disallowing extreme rotations, which can otherwise occur in some cases. Another possible type of movement constraint is to lock the rotation angle, disallowing movement by more than a certain angle away from the original normal of the piece.

The collection of all user specified constraints forms a complex design space that is difficult to manually navigate. Trying to satisfy one constraint may invalidate others. Optimization methods can deal with these complex constraint satisfaction problems effectively.

4 Implementation

Our design tool is implemented as a Rhinoceros 5 plugin using C#. It can be integrated into existing workflow with custom Rhinoceros commands and toolbar buttons, requiring little training. Rhino functions allow iterative building while optimizing for given hard constraints. The user can selectively remove and break constraints. More final modifications can be made as well: there exist functions to trim a planar piece before intersection with another piece, and to trim redundant parts of pieces based on a given silhouette (i.e. projecting the desired silhouette back on each piece and removing the resulting projection from all but one piece.) To prepare the output for printing, at any given intersection, slits are cut out of each piece at proper locations to enable construction. This is done naively, making sure that each piece can be slid into its partner, but the user must make sure that the construction can be assembled properly. We provide tools to aid in this process to identify where unbuildable cycles occur and either introduce an extra cut or rearrange the orientations of the slits according to user input. Finally, we perform rigid body simulations in Maya to check for stability of the final construct, as unwanted sliding in the slit direction can still occur.

Fig. 7 A dome structure exhibiting many variations in shape and orientation.

5 Examples

Figs. 1, 7, and 8 present examples of the process showing the possibilities of utilizing computation to handle constraints and enable unexpected structures. Note the diversity of constructions allowed by these constraints and the use of optimization methods to help in the process.

Fig. 8 A Shadow Art example (see [5]) with many more cycles. The example is constructed with silhouette constraints (the two shadows shown). In this case, the optimization is done as a post-rationalization step, and as there are many cycles, pieces tend to become more parallel.

6 Conclusions

As any project grows, the performance and construction constraints multiply. It is often impossible to foresee each issue from the initial stages of the design, and the complexity is such that the architect cannot account for every issue up front. This is a task that can be aided through the use of computation. A seemingly simple construction, orthogonally intersecting pieces, can quickly get complicated. We have presented an optimization-based design method that enables complex structures that are driven by their constraints while still allowing artistic freedom. As well as enabling a novel sculptural technique, since the results can be produced with a laser cutter, immediate applications of our method include prototyping and exploration of shapes, which can be of much use in the educational domain. Most importantly, by presenting a method of designing with the aid of optimization techniques to handle construction-aware constraints, we hope that this inspires further use of optimization during the design process rather than only as a post-rationalization step.

References

1. Bollinger, K., Grohmann, M., Tessmann, O.: The Sphere Project - Negotiate Geometrical Representations From Design to Production. In: Advances in Architectural Geometry, pp. 103–111 (2010)
2. Combettes, P.: Construction dun point fixe commun famille de contractions fermes. Comptes Rendus de lAcadmie des Sciences Srie I (1995)
3. Kilian, A., Nagakura, T.: Design exploration through bidirectional modeling of constraints. Massachusetts Institute of Technology, Cambridge (2006)
4. Mangelsdorf, W.: Structuring Strategies for Complex Geometries. Architectural Design 80(4), 40–45 (2010)
5. Mitra, N.J., Pauly, M.: Shadow art. ACM Transactions on Graphics 28(5), 156:1-156:7 (2009)
6. Oxman, N.: Get Real Towards Performance-Driven Computational Geometry. International Journal of Architectural Computing 5(4), 663–684 (2007)
7. Pottmann, H.: Architectural Geometry as Design Knowledge. Architectural Design 80(4), 72–77 (2010)

Modelling the Invisible

Benjamin Späth

Abstract. The paper describes shortly the system architecture of an immersive interactive virtual acoustics and visualisation system. Some realised models for education, research and building practice are presented. For educational purposes some existing concert halls like the Musikvereinssaal in Vienna or the Mozartsaal in the Liederhalle in Stuttgart are modelled by students to learn and percept directly material and geometrical impact on acoustics. For research purpose new concepts for acoustic spaces are explored in the real time environment. Due to new requirements deduced from new musical presentation forms represented e.g. by "spatial music" historical precedents are no longer helpful as prototypes. Lacking of archetypes the virtual acoustic system offers a method to evaluate novel spatial acoustic concepts by individual perception of acoustic and architectural aspects. Experiences and results made with the modelling of the "Neue Stadthalle Reutlingen" design by the architect Max Dudler are presented as an example for the systems practical use. Due to sensitive public perception of cultural buildings like the mentioned case study the individual perceptible demonstration of the future hall increases the acceptance of the planning. The virtual acoustic-visual system offers even laymen to get an idea of how the future hall will sound. Concluding the presented experiments underlines the importance of modelling not only visual aspects of architecture but also invisible features like room acoustics as an integrated perceptible and interactive representation of a multi aspect architectural design.

1 Introduction

Concert halls make high demands not only on aesthetics but also on acoustical aspects of design. The challenge for designers is the integrative development of both aesthetic and acoustic quality. As acoustic spaces like concert halls are rooms of

Benjamin Späth
University of Stuttgart, Germany

perception and the subject of room acoustics itself is very abstract and complex, it is important to provide an integrated representation of perceptive aspects of the acoustic design, this is visualisation and auralisation of the design. Tools for room acoustical simulations like Odeon [5], Catt Acoustics [6] or EASE [4] already exist. Also for the visual simulation of architecture tools like COVISE/COVER [9] are available. Figure 1 shows the resulting diagrams from acoustic engineer like simulations.

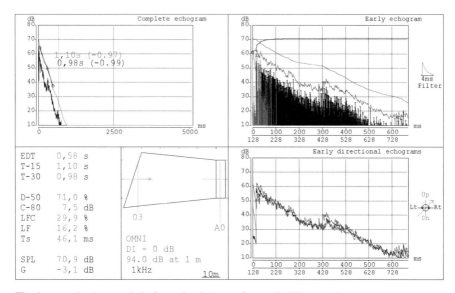

Fig. 1 acoustic characteristic from simulation software CATT acoustics

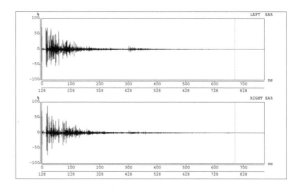

Fig. 2 corresponding impulse response representing all relevant acoustic information at a specified frequency

The acoustic simulations are very precise. But the diagrams and figures which result from these simulations are very abstract so that mostly engineers are able to interpret them. Even so a simultaneous perception of visual and acoustic aspects is not provided in a sufficient manner. But a simultaneous perception of both aspects, acoustic and visual, is provided by a system combining an immersive interactive virtual environment with a real time room acoustical simulation [8].

2 System Architecture

The system is a combination of a collaborative interactive real time virtual environment called COVISE/COVER [9] and a real time room acoustics simulation called RAVEN/VA [7]. The core intention is to integrate the advantages of both systems into a single system which is feasible in an architectural design context. Due to computational power consuming acoustic simulations the resolution of the acoustic system is limited to q a few tens of polygons whereas the visual system (COVER) which is using a pre-calculation method of light information by embedding it into a texture map handles several ten thousands polygons in real time. Furthermore due to different data formats and different geometrical resolutions of the systems two different models are used.

Figure 3 shows the scheme of the client side integration of the simulation into the COVER architecture. The visualisation and the acoustical simulation module are communicating through TCP protocol on a local network. The user position and interaction data are distributed to both applications by the tracker daemon. Thus both applications calculate the corresponding output although they are using different models. This requires certainly that the models are persistent in terms of origin, scale and orientation.

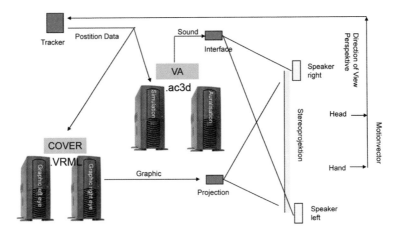

Fig. 3 scheme of the client side integration VA/RAVEN and COVISE/COVER (Spaeth 2008, 371)

As mentioned above the two simulation systems are based on different data formats. While the visual system COVISE/COVER is based on the ISO standard

VRML 97 and X3D the acoustic system is based on a proprietary format ac3d [2]. The COVER/COVISE systems supports not only VMRL 97 standard nodes for interaction or geometry description, but also provides extended specialised features that allows e.g. to generate or move objects in the model during runtime. These features remain active when combined with the acoustic system. The simulation of the acoustics is a twofold process. Based on material assignments to the geometry the impulse response of the current position is calculated. Then this binaural impulse response is convoluted onto an anechoic input signal to a stereo output signal which can be provided by headphones or loudspeakers.

To include the creation of the acoustic models into an architects work flow it is necessary to provide transformation tools between common CAD formats to required ac format. The application White_Dune [1] was adapted to transform VMRL files which are the base also for the visual model into required ac files. Furthermore a movable sound source is implemented into the visual acoustic model.

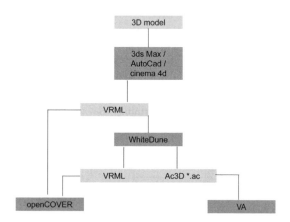

Fig. 4 flowchart model creation

3 Education

The individual perception of different acoustic spaces helps students to understand the meaning of the specific acoustic characteristics. Through the combination of visual and acoustic perception the relation between architectural space and acoustic characteristics becomes more obvious to the students. Thus the students are asked to create models of well-known concert halls or operas like the "Golden Hall" in Vienna or the "Mozart Hall" in Stuttgart. Due to the different resolution of the acoustic and visual model the students are forced to simplify the acoustic model. So complex geometry and material properties have to be interpreted and transformed into reduced geometry with adapted material properties. Through this cognitive process

 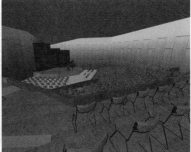

Fig. 5 students model of the "Golden Hall" Musikvereinsaal in Vienna

Fig. 6 students model of the Mozart hall in Stuttgart

of transformation the impact of material and geometry to room acoustics is realised. As architects are strongly related to the visual representation sophisticated textures with pre-calculated lighting information is provided.

4 Research

Due to changed interpretation practices of contemporary music like Stockhausens spatial piece Gruppen Nr.6 [3] composed for three orchestra groups traditional precedents of concert halls are limited in being prototypes for spatial concepts for acoustic spaces. Thus methods are needed which have the capacities to explore new spatial concepts by perceiving the design individually as an architectural representation including visual and acoustic aspects. Using the described system the designer is able to make the design perceivable to different users. Thus one could document the impact of the spatial configuration of the enclosure in relation to the spectators and the musical composition by questioning the individual impressions. By an inter-subjectivity process universal statements about the relation between architectural concept and acoustic perception could be made. This research is to be done and here the author wants to give an idea about what kind of research is conceivable.

At the moment the system is applied for exploration, presentation and evaluation of acoustic concepts where historic precedents are no longer useful, as mentioned above.

The diploma work which is presented is based on three potential orchestra positions. The main architectural concept is to link these orchestra positions together. Circuits allow the players to traverse through the spectators reaching the different orchestra positions. With the interaction possibilities of the system this moving orchestra can be simulated on an abstract level. Thus the spatial and acoustic concept of the design can be evaluated precisely by perceiving the idea tangible on the system.

Fig. 7 diploma "concert hall for spatial music" by Maryem Cengis

5 Practice

For the design of the "Neue Stadthalle Reutlingen" a high quality concert hall was required by the client. The public discussion about the project is controversial. Thus the county administration wants to present the future hall to the interested public to influence the public opinion. In cooperation with the acoustic engineers BeSB GmbH Berlin and the architect Max Dudler an interactive acoustic model of the concert hall was created. The core aim of the model was not to proof the acoustic concept of the engineers but to convince the participants and users about the quality of the design. Most of the decision makers are laymen, which are not able to interpret the abstract diagrams and figures from the engineers. Thus the individual perception of the hall in the presented system, walking freely in the model, testing and exploring preferable seats, enjoying the view from the stage by synchronously hearing the real acoustic of the hall gives the participants an individual impression and base for their decisions.

As mentioned the system is not intended to substitute calculations by engineers, but it is intended to make the results of these calculations perceivable directly.

Fig. 8 interior view visualization from the "Neue Stadthalle Reutlingen", provided by Max Dudler Architekten Berlin

Fig. 9 interactive acoustic model "Neue Stadthalle Reutlingen"

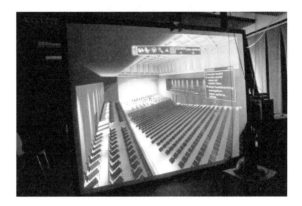

Fig. 10 interactive acoustic model "Neue Stadthalle Reutlingen"

6 Conclusions

The paper describes roughly a virtual acoustic environment composed of a high resolution interactive visual simulation and a very precise acoustical simulation. It presents three different applications of a virtual acoustic environment. Besides the academic fields of education and research the system demonstrated its ability in a practical context. It turns out as a useful tool for the transmittance of acoustic concepts by modelling and interpreting existing acoustic spaces. It gives proof of being an evaluation tool for reviewing acoustic concepts which are missing comparable precedents. Applied to the architectural practice the system demonstrated its ability to transform abstract acoustic requirements into a perceivable level which is accessible even for laymen. Thus different decision makers can participate on the acoustic architectural concept.

Individual perception of architectural concepts and designs concerning spatial and acoustic aspects is the focus of the system. It is not intended to substitute acoustic engineering. Contrariwise it helps to support the engineer like aspects of the design. The invisible aspects of the design can be integrated into the design and the decision process. The perception of spatial enclosure is not limited to the visual aspect but is now extended to the acoustics. The presented system and the integration into the common modelling process of architects by the developed tools hold a high potential to integrate the invisible into the architectural consciousness.

References

1. White_Dune - graphical VRML97 Editor and animation tool (2008), http://vrml.cip.ica.uni-stuttgart.de/dune/ (accessed May 30, 2011)
2. AC3D: 3D Design Software: inivis limited (2010a), http://www.inivis.com/ (accessed May 30, 2011)
3. Karlheinz Stockhausen Wikipedia (2010b), http://de.wikipedia.org/wiki/Karlheinz_Stockhausen (accessed November 17, 2010)
4. Ahnert, W.: EASE: SDA Software Design Ahnert (2010), http://www.ada-acousticdesign.de/set/setsoft.html
5. Christensen, C.L., Rindel, J.H.: Odeon A/S, Odeon (2010), www.odeon.dk
6. Dahlenbäck, B.-I.: CATT-Acoustic. Gothenburg (1998), http://www.catt.se
7. Lentz, T., Schröder, D., Vorländer, M., Assenmacher, I.: Virtual Reality System with Integrated Sound Field Simulation and Reproduction. EURASIP Journal on Advances in Signal Processing (2007)
8. Spaeth, A.B.: Room acoustics in the architectural design process. In: Muylle, M. (ed.) Proceedings of the 26th eCAADe Conference on Education and Research in Computer Aided Architecutral Design in Europe, pp. 367–374 (2008)
9. Wössner, U.: Virtuelle und Hybride Prototypen in Kooperativen Arbeitsumgebungen. Dissertation, Universität Stuttgart (2009)

Ornate Screens – Digital Fabrication

Daniel Baerlecken, Judith Reitz, Arne Künstler, and Martin Manegold

1 Introduction

The English word ornament comes from latin 'ornamentum', rooted in 'ornare', which can be translated with 'to give grace to something'. In the past centuries diverse discussions about philosophy and necessity of ornament have formed. Alberti states in *De re aedificatoria* (1485) that a building is developed through massing and structure first; ornament is added afterwards to give the bold massing and structure *'pulcritudo'* (beauty) as the noblest and most necessary attribute. Ornamentation serves to increase the *'pulcritudo'* of a building and to create harmony in the overall intellectual and primary framework. A few hundred years later the critical theories of *Adolf Loos* (1908) describe ornament as immoral and degenerate –a suppression of the modern society neglecting the modern technologies.

Afterwards ornament has been abolished by the avant-garde and has been replaced by material aesthetic. Recent discussion overthrow both - the Albertian and the Loos' Paradigm. Ornament is not anymore a parergon, an unnecessary and decorative accessory. The parergon, the ornament, actually becomes the ergon, the main element: ornament becomes performative . *Strehlke and Loveridge* (2005) underline this assumption in their essay about "the Redefinition of Ornament". The paradigm shift has been in the architectural discourse from modern architecture to digital architecture.

Daniel Baerlecken
Georgia Institute of Technology, Atlanta, USA

Judith Reitz
RWTH Aachen University, Aachen, Germany

Arne Künstler · Martin Manegold
Imagine Structure, Frankfurt, Germany

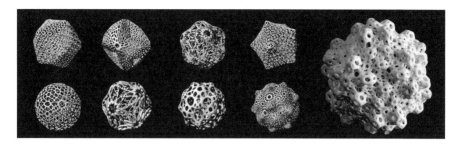

Fig. 1 Lights series 2010, 1-9

As ornament has been eliminated in modern architecture by mass production, today's Computer Aided Architectural Design (CAAD) and Computer Aided Architectural Manufacturing (CAAM) allow the re-introduction of ornamentation and variation. In this way modern architecture's mass production is replaced with mass customization and allows the re-introduction of variation.

2 Perceptive Performance vs. Structural Performance

How does ornament perform in a perceptual way? Ernst Gombrich describes in the *sense of order (1979)* the guidance by the aesthetic enjoyment as an active agent reaching out toward the environment. Ornamentation allows "Einfuehlung" - the identification with an object. By using variation and the idea of the mass customization ornament can become performative combining the 'Einfuehlung' (empathy) as perceptional way with the constructional aspect.

Lars Spuybroek writes in 'the architecture of continuity (2008)':[…] No such distinction is made in the Gothic; there, the structure itself is 'vitalized'.[…]all forms are results of movements, structural movements that are shared, that are passed from one to the other. What is crucial to note, though, is that movements and variations are not just tools of deformation: they are also structuring states of singularities – that is, transformative properties.[…] Exactly the structural fitness of ornaments needs to be combined with the aesthetic enjoyment to generate a new performance and repose.

The performative tool is here the feedback between the engineering and design software, while both inform each other. The three following experimental projects are based on the systems that explore structure and ornament as one entity in an anti-Albertian approach. Ornament becomes structural, the decorative aspect becomes necessity.

3 Prefabricated Light Series

The first example is the design of a parametric series of lights (see Fig. 1), which are fabricated through STL and SLS technologies.

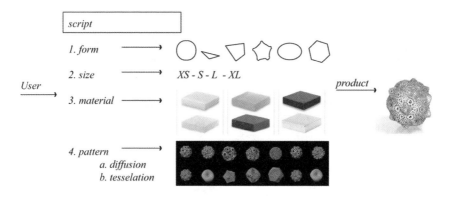

Fig. 2 Design and production process

The objects are brought to life primarily as single objects without joints or seams as 'additive fabrication'. By that, the objects can be produced directly from digital information as unique pieces in diverse materials from alumide, steel, glass, plastic or ceramic. The size, material, form and pattern vary through the order of the user.

The choice of material varies automatically the thickness of each lamp. With plastic the thickness of the material can go down to 1mm since the form of the lamp enhances stiffness and stability. Each light can be different from the other, but still belong to the same family. Changing variables in the original script allows creating a different light, which still can feed back to a greater family. One can create unique objects by manipulating the four parameters. Eventually the series defines a particular number of ten light weight structures, which aim to minimize the use of material by use of ten different patterns informing the objects stability, size and thickness.

Similar customized projects are successfully explored by Mi-Addias shoes, Freitag bags or in the area of customized Asian designer clothes copies. Mass production is replaced with mass customization; each user can order and identify with its own individual object.

4 Exhibition Installations

The second experiment is based on two exhibition installations. The DS 2010 project was designed and built for the 'Designers Saturday 2010' in Switzerland. The second installation will be exhibited at the 'Passagen 2012' in Cologne.

The DS 2010 project will be discussed in so far as it informs the Passages 2012 project.

Fig. 3 Exhibition at the DS 2010 in Langenthal, Switzerland

The DS 2010 project focuses on tessellation to create aesthetic enjoyment, whereas the Passages 2012 project studies structural aspects: perceptive performance vs. structural performance.

4.1 Perceptive Performance

The DS installation is a wall that twists to create a passage through the installation. The project uses a surface tessellation that is directly reacting to surface curvature. The polygonal tessellation transforms in order to react to changes in curvature by scaling the size of the module and by reconfiguring the typology (notes within the polygonal structure vary through their local relations). The installation is constructed out of plywood with a joint system. All elements are different and unique. Therefore one of the main challenges was to automate the numbering of the parts so that the wall could be assembled easily. The installation was fabricated through water jet technology, even though wood is not ideal for water jet cutting. The machine had to be modified to use as little water as possible to void damage to the plywood, which was successful.

For the first project no structural calculations were made due to a narrow time frame for design and fabrication. The structure had to be tested and corrected during the erection of the structure. Certain problems were not foreseen and had to be fixed through suspension cables.

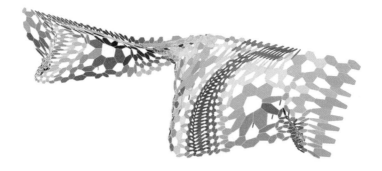

Fig. 4 Size analysis DS 2010

The tessellation with varying parts was able to create an interesting aesthetic experience, but had deficits in structural and constructive performance. The first problem was resulting from parts that were too small. Small parts (Fig 4: pink parts) were weaker than medium scale and large scale parts, because the small size of the panels only allowed for a small connection element, which weakened the structure substantially. The second problem was based on the decision to introduce a large variety of panel shapes from equilateral panels to elongated panels. The elongated panels created more tension in the connection pieces, which caused a few connections to fail.

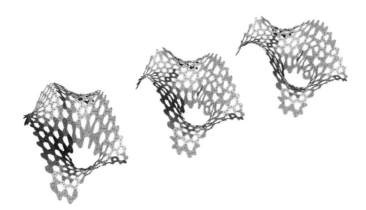

Fig. 5 Optimization of the structural model from left to right

4.2 Structural Performance

Design methodology: In order to avoid problems of the same sort, the second exhibition at the Passages 2011 tries to integrate structural design and spatial design at the same time. Therefore a link between architects and engineers had to be defined at the beginning.

This link is described in the following, even though the project might still change. The exhibition proposal is also different formally. Instead of a wall the second project proposes a structure of arches that join in a cone.

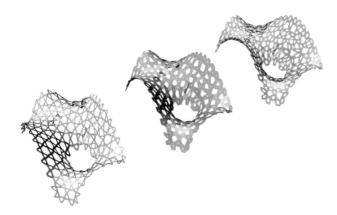

Fig. 6 Optimization of the structural model from left to right

Fig. 6 shows a series of optimizations. The model on the left shows the first tessellation. The tessellation follows a Nurbs surface, but the deviation from the surface curvature is not controlling the tessellation. The tessellation follows a structured grid. The panels have almost the same size: pink indicates small panels and blue large panels. But there are still two problems. Firstly the skin appears quite facetted in some areas. Secondly the blue panels that surround the cone tend to be elongated panels. The size of the pink of panels is already big enough to avoid the problem of weak connections. The next iteration, diagram in the middle, shows the model after a surface relaxation procedure: in general the surface is smoother and we can observe that the relaxation algorithm has produced more equilateral panels. The last iteration tries to create even more equilateral panels.

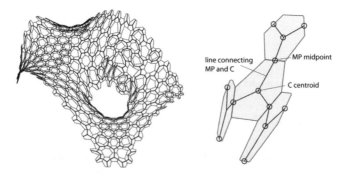

Fig. 7 Geometric information exchanged between Rhinoceros and R-Stab

In the design process exchanging structural data between architects and engineer became essential. The architects prepared a model that contained the centroids, the midpoints of each polygon edge that was connecting to another panel and lines that were connecting these points.

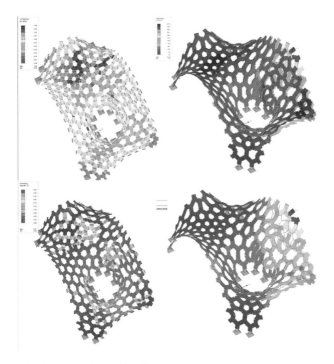

Fig. 8 Stress ratio and bending moment

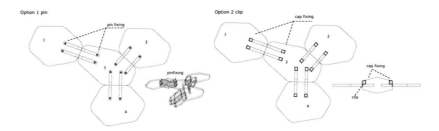

Fig. 9 Variation of the module

Currently we are studying ways to modify the module in such a way that each module is changing according to its structural performance: the module thickens where it is necessary and gets thin in areas with lower structural performance.

This we can achieve varying the outer perimeter of the hexagonal panel and changing the inner opening of the panel.

In order to automate this process a link between Grasshopper and R-Stab has been established to study the results in real time. Angles analysis is automated so that feedback can be given almost in real time.

5 Outlook

We are currently building mock-ups with different connection details and we are looking at alternative materials to plywood. But it seems likely that we will use the same material, but use cnc-milling instead of water jet cutting this time. We might introduce different board thicknesses to work with the stress ratio. Also we continue the study between Rstab and Rhinoceros to ensure fluid transition of our data.

We see a high potential to introduce tools that facilitate collaboration between architects and engineers in an early stage for future projects. We think that the 19th century split between engineers and architects can be re-thought through collaborative tools that allow two-directional feedback.

References

Alberti, L.: De re aedificatoria, on the art of building in ten books (translated by Rykwert, J., Leach, N., and Tavernor, R.). MIT Press, Cambridge (1988)

Bloomer, K.: The nature of ornament. rhythm and metamorphosis in architecture. Norton & Company, New York (2000)

Gell, A.: Art and Agency – An Anthropological Theory. Oxford University Press, London (1998)

Gombrich, E.: The sense of order – a study in the Psychology of Decorative Art. Phaidon, London (1979)

Spuybroek, L.: The aesthetics of variation. In: The Architecture of Continuity, Nai Publisher, Rotterdam (2008)

Strehlke, K., Loveridge, R.: The Redefinition of Ornament, Computer Aided Architectural Design Futures 2005. In: Proceedings of the 11th International Conference on Computer Aided Architectural Design Futures, Vienna, pp. 373–382 (2005)

Worringer, W.: Abstraction and Empathy: A Contribution to the Psychology of Style. Elephant Paperbacks, Chicago (1908)

Künstler, A., Manegold, M., Reitz, J., BAerlecken, D.: Digital Processes. Detail Magazine (2011)

The Railway Station "Stuttgart 21"
Structural Modelling and Fabrication of Double Curved Concrete Surfaces

Lucio Blandini, Albert Schuster, and Werner Sobek

Abstract. Contemporary architecture often asks for complex double curved surfaces while the engineering of such structures often becomes a challenge. This paper illustrates the tasks that emerged and the methods developed during the engineering process of the new railway station "Stuttgart 21". One of the main challenges in generating a structural model out of the architectural design was the definition of the middle surface for the concrete shells and the need to obtain a meshed system with continuously varying shell thicknesses. Both tasks were solved using RhinoScript, which was also used in the further design stages to define the rebar cage on the basis of a parameter-based optimized design. Fabrication costs were thereby minimized by reducing the number of reinforcement types throughout the structure.

1 Introduction

It is often surprising how light and elegant concrete can appear once used for spatially curved surfaces. However the engineering of such complex structural elements is an extremely challenging task, which requires close cooperation between the different professionals involved. The structural engineer should not only master the understanding of the structural behaviour of shells and the material properties of concrete, but should also be capable to develop adequate computational tools to optimize the structural behaviour of the design.

Several computational tools and fabrication concepts have been developed at Werner Sobek Stuttgart over the last ten years in order to support the engineering and fabrication of complex double curved concrete structures. At the Lufthansa

---

Lucio Blandini · Albert Schuster · Werner Sobek
Werner Sobek Group

Aviation Centre, ten modular "fingers" are roofed with double-curved reinforced concrete shells. Each roof module is different in geometry but all of them have been derived parametrically in AutoCAD from the same typology. The shells are of a constant thickness and are stiffened by means of binding beams.

While the shells are geometrically slightly different, the repetition of curvature radii in the structural elements allowed for a high reuse rate of the formwork, thus keeping the costs within a reasonable range. The moderate curvature of the radii allow for a high number of rebars to be bent directly on site.

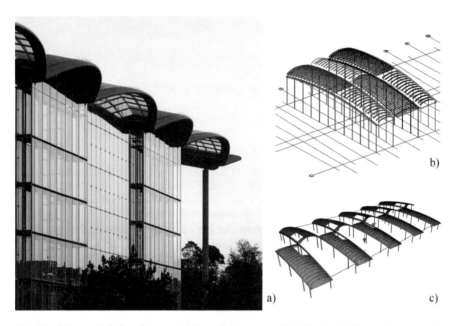

Fig. 1 Lufthansa Aviation Centre. a) View of the corner (H.G. Esch) b) Geometrical model (Werner Sobek Stuttgart) c) Structural model (Werner Sobek Stuttgart)

In the new Mercedes-Benz Museum, Stuttgart, the most complex concrete elements, the "ramp" and the "twist" were engineered as curved hollow box girders with varying sections (Fig. 2c). In order to determine the concrete sections, the middle surfaces of the shells were derived based on the original outer surfaces, which were imported from the architectural Rhino model. The middle surface model was then meshed, analysed and structurally optimized with the FEM-program Sofistik. The thickness of the concrete walls was kept constant to 50 cm. The formwork for the double-curved surfaces was prefabricated and costs were again minimized by the reuse of repeating elements. To maintain the correspondence between the geometric information in the 3D-model and the built surfaces, a cloud of reference points was defined, to allow on-site verification of the respective coordinates.

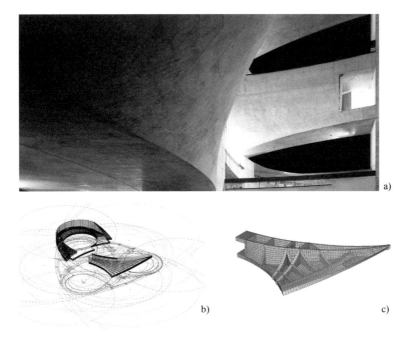

Fig. 2 Mercedes-Benz Museum. a) View from inside (Brigida Gonzalez) b) Geometrical analysis (UN studio) c) Twist structural model (Werner Sobek Stuttgart)

The modelling of the railway station "Stuttgart 21" took advantage of the experience gained in the two described projects. However the project was in many ways more challenging because of the constantly varying concrete shell thickness. New tools had to be developed in order to precisely define the thicknesses at each node of the finite element mesh. Moreover, the radii of curvature were much smaller here compared to the Lufthansa Aviation Center and the Mercedes-Benz Museum, which requires a considerable effort in the definition of the rebar geometry. The bars must be bent off site, and while most of them are one-directional, a certain portion must be bent in two directions. The structural design and the tools created for this process are described in the following paragraphs.

2 Stuttgart 21: Design and Geometry

The railway station "Stuttgart 21" is the core project in an overall effort to modernize the connection between Munich and Paris. The present terminus station in Stuttgart will be replaced by a 450 m long and 80 m wide underground through station designed by Ingenhoven Architects of Düsseldorf. The space is characterized by 28 doubly curved concrete columns; the so called "chalices".

The railway station was modelled architecturally using Bentley Microstation. The top surface of the station has five different slopes. The platform level has two different slopes in the longitudinal axis. These boundary conditions require different heights of nearly each chalice. Nevertheless a concept was implemented to achieve a maximum amount of geometrically common elements in the concrete shell roof.

Fig. 3 Image of physical model (Werner Sobek Stuttgart)

The geometric description of the doubly curved roof has been split at level -6m (Fig. 4). This allows the upper part of the standard chalices to be described in a way that more or less follows one type of geometry. The different heights at each chalice are mitigated by three types of pedestals which connect the lower end of the chalices with the inclined platforms. Nevertheless each pedestal has to be adapted to the height of each chalice, by sectioning the pedestal at the upper surface of the platform. Through this approach, a high number of geometrically similar chalices were achieved: 23 so called "standard-chalices", 4 "flat-chalices" and one special chalice. Still, several chalices present local changes as they are partly cut at the junction with the sidewalls and at the kinks of the roof. The grouping of the pedestals also leads to a high number of identical surfaces, as they only differ in the lower portion.

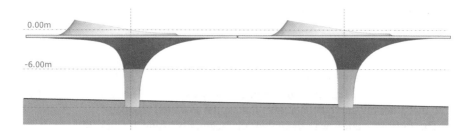

Fig. 4 Section showing the division line for the chalices (Werner Sobek Stuttgart)

3 Structural Modelling

The key in setting up a structural model that comprised so many variations of the same typology was the automation of generating the finite element meshes. This was achieved directly in Rhinoceros through scripting. The meshed model contains all the information necessary for the finite element analysis and the data could be exported directly from Rhinoceros into Sofistik by means of text instructions. Sofistik was then used as the structural analysis software.

The structure was divided into areas which could be better modelled by means of beam elements (lower area of the chalices) and areas which had to be modelled by means of shell elements (upper area of the chalices). The interface between these areas was modelled and checked carefully to assure a correct force transfer. The differentiation led to the coding of two sets of scripts to derive the model information.

One of the key issues for both sets of scripts was the geometrical definition of the centre axis respectively the middle surface. The centre axis was modelled by sectioning the surfaces which describe the lower area of the chalices and by connecting the gravity centres of the sections. The axis geometry and the sections were automatically exported into a text instruction file for further use in Sofistik.

The generation of the middle surface of the shell-meshed regions was more complex. The outer and inner surfaces were pre-processed, defining different sub-regions to be later analysed in detail in Sofistik. These surfaces constitute the input for a script which determines the shortest distance between the two surfaces at a discrete number of nodes and within a certain range of approximation. All the necessary information (i.e. element numbering, node thicknesses, etc.) was exported into a text file for further use in Sofistik.

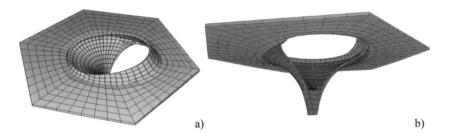

Fig. 5 Structural model of a chalice. a) View from above b) View from below (Werner Sobek Stuttgart)

This process was developed and checked for one single chalice first, and then further implemented to model the whole station, thereby accounting for all the local variations from the standard geometries. Due to the size of the meshed model, the elements were sorted in groups to better address structural evaluations, rebar topology, construction phases, and other issues. The main rebar directions were also directly defined through scripts in Rhinoceros.

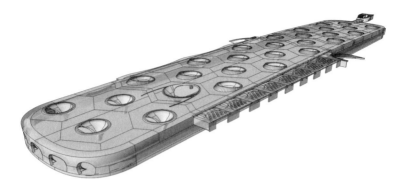

Fig. 6 Architectural model of the railway station in Rhinoceros (Werner Sobek Stuttgart)

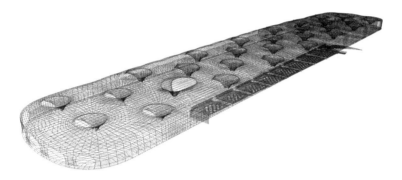

Fig. 7 Meshed model of the railway station in Rhinoceros, middle surface (Werner Sobek Stuttgart)

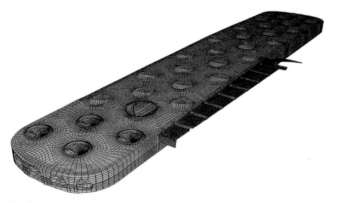

Fig. 8 Finite element model of the railway station in Sofistik (Werner Sobek Stuttgart)

For the Stuttgart 21 project, the checking of the model information becomes even more important than usual, as the workflow is highly automated. For this reason, the correspondence between the given architectural geometry and the finite element model was checked visually by sectioning every chalice 24 times, as well as analytically by means of data comparison at special sampling nodes. The overall process of structural modelling was carried out iteratively multiple times to optimize the structure.

4 Fabrication Issues

The architectural 3D-model is the basis for the fabrication of the formwork, with scripting used to define sample points on the surfaces of the formwork for on-site geometric control. The generated sample points and coordinates were inserted in control-point drawings. For the area at the special chalice two sets of control-point drawings had to be generated. The structure in this area has to be precambered, therefore one set of control-point drawings considers the deflections under dead loads, the second set does not.

The definition and optimization of the rebar geometries was a very complex task, due to the small radii of curvature of the chalice-geometry. The architectural 3D-model was split and tailored into different sets of surfaces to generate the rebars. This allows for different reinforcement strategies in consideration of geometric and structural issues. The rebar layout was generated mainly through scripting, defined through sets of parameters such as distance to the surface, anchoring length, and rebar distance.

Fig. 9 3D-model of the rebars cage produced parametrically (Werner Sobek Stuttgart)

As the bars are generated based on doubly curved surfaces, the resulting geometries are NURBS splines. Scripts were developed to convert these splines into polylines made of straight segments or arch segments with similar rebar geometries grouped together within certain tolerances. Fabrication issues such as segment lengths and the deviation between the original and converted geometries were also verified using computer scripts, thus allowing for further optimization and control.

5 Conclusions

The experience gained over the last 10 years with structures like the Lufthansa Aviation Center and the Mercedes-Benz Museum was the basis for the structural design process of the High Speed Railway Station "Stuttgart 21". This design process has been set up with a partly automated workflow, making use of specially developed algorithms to allow an iterative structural optimization.

During the design development phase, the main challenge was the generation of a detailed finite element model with constantly varying thickness based on the architectural 3D-model. Later on, during the preparation of the construction documents, the definition of curved or polygonal rebars, matching the steady change of given surface curvatures, was particularly challenging. In this case, it was fundamental to consider the rebar production limits and strive to keep the costs within a reasonable range. Overall, the use of scripting has proven to be a very powerful tool by allowing automated changes in the modelling work and in optimizing the workflow between the different professionals involved.

Acknowledgments. The engineering of such a complex project as "Stuttgart 21" would not be possible without a skilled team. The authors wish to specially thank all the managing engineers who have been in charge of coordinating and supervising the whole team work: Dietmar Klein, Jan Ploch, Roland Bechmann, Torsten Noack, and Steffen Feirabend. Thanks are also due to Matthias Rippmann, Silvan Oesterle, and Michael Knauss for their contribution to the scripting work.

References

Sobek, W., Reinke, H.G., Berger, T.: Die neuen Verwaltungsgebäude der Deutschen Lufthansa AG in Frankfurt am Main (Lufthansa Aviation Center). Beton und Stahlbetonbau 1, 1–4 (2005)

Sobek, W., Klein, D., Winterstetter, T.: Hochkomplexe Geometrie. Das neue Mercedes-Benz-Museum in Stuttgart. Beratende Ingenieure 10, 16–21 (2005)

Performative Surfaces
Computational Form Finding Processes for the Inclusion of Detail in the Surface Condition

Matias del Campo and Sandra Manninger

Fig. 1 Austrian Pavilion Shanghai Expo 2010 South Façade, SPAN & Zeytinoglu, photo © Maria Ziegelböck

Abstract. The main aim of this paper is to explore the relationship between form finding processes and the simultaneous emergence of detail solutions in the process, in opposition to concepts of addition. Additive systems form the prevailing method in the discipline of architecture when it comes to solve problems of transitions between interior and exterior. Another convention in architectural design is the usage of detail in the moments of transition between varying material conditions. It is still a question of definition as to what can be considered a detail, provided the conversation focuses on aspects which form an alternative mode of thinking in the architectural realm. With the aid of the example of the Austrian Pavilion for the Shanghai Expo 2010, an optional method in the relation between surface behavior, performance and detail is examined in this paper.

Matias del Campo,
Sandra Manninger
SPAN, Dessau

1 Conventions in Modes of Surface Formation

The main aim of this paper is to explore the relationship between form finding processes and the simultaneous emergence of detail solutions in the process, in opposition to concepts of addition. Additive systems form the prevailing method in the discipline of architecture when it comes to solve problems of transitions between interior and exterior. Another convention in architectural design is the usage of detail in the moments of transition between varying material conditions. It is still a question of definition as to what can be considered a detail, provided the conversation focuses on aspects which form an alternative mode of thinking in the architectural realm. The conventions of the definition gravitate around two possible modes of explanation, on the one side leaning towards the description of the transition from one architectural element to another, such as the transition from the glazing of the window to the wall by the use of frame. Another ecology of explanation unfolds by defining the detail as a constituent mean of expression. The examples of this plane of thought are multiplicious: from the Paxton Chrystal Palace to the Centre Pompidou to the Hong Kong Shanghai Bank of China building by Foster & Partners, where the detail establishes, beyond its purely functional role, as specific moment of embellishment, a mechanical ornamentation. Of course the concept of detail as an additive system is closely related to the means of fabrication available to architects throughout history, which explains the prevailing presence of tectonics in architectural production. However it is also fair to speculate about the cultural relevance of the concept of addition in relation to the respective era. Does the architecture of a specific era reflect back insights of the age? To illustrate this notion we can rely on one example from the Baroque era: the human body as a mechanism, like a mechanic watch or an Automaton[1]. The Universe, the gravitating planets, and their trajectories around the sun where observed like a mechanical instrument. A composition of wheels and cogs, a collection of distinctive parts in constant, balanced interplay. Does the architecture of the respective age reflect this notion too? It is a known fact that artists and architects, such as Borromini and Bernini worked together with mathematicians in order to understand specific geometric problems in their architecture. (Such as Borromini collaborating with Virgilio Spada at the Villa Pamphilij Rome[2]...) If we follow this line of thought it would suggest the notion in architecture to create detail is based on a global view on the composition of the world. A mechanical system thus would suggest the use of individual, defined components, such as wall, frame and glass for the fenestration.

2 Definition

Detail: In terms of a prevailing architectural design practice, the design of a building can be divided into specific areas: The preliminary design that in most cases does not involve the design of details, the design process that includes details on a sketchy level and the drafting of plans for the construction site, which includes all the details for the construction crew and for prefabrication of details. This paper discusses a novel, alternative path in the lineage of architectural design, whereas the detail is not considered an individual component in the process but in fact becomes an integral part of architectural body. This novel definition incorporates the task of the detail in the surface. The detail can be considered a gradient change in the surface condition. Manuel de Landa describes the possibility to control the quality of material in a gradient fashion in his essay *Uniformity and Variability:*

One aspect of the definition of the machinic phylum is of special interest to our discussion of contemporary materials science. Not only is the phylum defined in dynamic terms (that is, as matter in motion) but also as "matter in continuous variation". Indeed, these philosophers define the term "machinic" precisely as the process through which structures can be created by bringing together heterogenous materials, that is, by articulating the diverse as such, without homogenization. In other words, the emphasis here is not only on the spontaneous generation of form, but on the fact that this morphogenetic potential is best expressed not by the simple and uniform behaviour of materials, but by their complex and variable behaviour. In this sense, contemporary industrial metals, such as mild steel, may not be the best illustration of this new philosophical conception of matter. While naturally ocurring metals contain all kinds of impurities that change their mechanical behavior in different ways, steel and other industrial metals have undergone in the last two hundred years an intense process of uniformation and homogenization in both their chemical composition and their physical structure. The rationale behind this process was partly based on questions of reliability and quality control, but it had also a social component: both human workers and the materials they used needed to be disciplined and their behavior made predictable. Only then the full efficiencies and economies of scale of mass production techniques could be realized. But this homogenization also affected the engineers that designed structures using this well disciplined materials[3.]

The authors of this paper would like to expand the before mentioned thought which deals with the properties of material and their structural qualities to the field of detail in architecture. In fact the interplay between structure, surface and detail create an interconnected lattice of forces and behavior, capable of describing volume and enclosure in an alternative fashion to the prevailing techniques of tectonics in architecture. In the case of this paper however the focus will be on the performative quality of the surface condition and the solutions applied to achieve the performance of detail in the skin of the Austrian Pavilion for the Shanghai Expo 2010. Skin, in this conversation is understood as a part of the surface and thus as a part of the architectural definition of the pavilion.

3 Surface

The general definition of surface relies on the idea of the surface being the uppermost boundary of an object or the boundary of a three dimensional object. In the framework of the discussed issues in this paper the definition of surface relies on the mathematics of topology, thus a plane of thinking in line with the description of topological surfaces by Gauss, as observed by mathematician and historian Morris Klein: By getting rid of the global embedding space and dealing with the surface through its own local properties "Gauss advanced the totally new concept that a surface is a space in itself".[4]

4 Precursors

The idea to embed the performance of details into the form of the surface first emerge in the work of the authors in the design for the traveling exhibition "Housing in Vienna" an exhibition commissioned by the Az W the Architecture Center Vienna. The demands of the commission included issues such as lightweightness, ease of transportation, ease of assembly and cost effectiveness. In order to achieve all these demands the decision was made to use a tessellated surface as point of origin. A series of pods was designed, and produced, based on the mathematical

Fig. 2 Housing in Vienna exhibition, © SPAN 2007

Fig. 3 assembly of pod - stoma

figuration of a Cairo tessellation. More important than the basic figuration was the discovery to embed details in the surface form instead of adding more elements to the pods, and thus reducing the setup time. In the case of the pods for the exhibition "Housing in Vienna" the embedded detail comprised of a mound-shaped protuberance that nested the perforations of the surface that house the zip-ties, the connection between the surfaces to hold them together. The reason for those mound-shaped protuberances lies in the necessity to distribute the stress forces implied by the zip-ties on the thin surfaces made of vacuum-formed PET Sheets. The intention was to reduce the thickness necessary to form the parts of the pod whilst achieving strong stability by the means of form. In a prevailing method, the solution would have been to add washers between the zip-ties and the surface, thus adding an additional element to the number of pieces that has to be put together to compose one pod. To avoid the increased number of elements, the approach to embed the detail in the morphology of the surface was chosen, and thus the setup time and the ease of handling increased dramatically. In preliminary tests conducted on the premises of the Az W in Vienna one person needed two minutes to setup one pod, including the lighting system. In general the entire exhibition, comprising of thirty pods and twenty wall panels can be setup by three persons in one day. The desire to reduce the amount of parts was already explored in the precursor of the housing exhibition, the exhibition design "The Austrian Winery Boom" also commissioned by the Az W. The authors where interested in the possibility to reduce elements by the inclusion of program into the surface condition. Doing so results in an elegant and simple way of connecting components as well as reducing the steps of fabrication and setup, whilst achieving a continuous gesture in the formation of aggregations. The main characteristic of both examples, the exhibition design "Housing in Vienna" as well as the exhibition design "The Austrian Winery Boom" is the use of tessellations in combination with topological surfaces as main method of design. These characteristics of the morphological process form also the basic design principles of the Austrian Pavilion for the Shanghai Expo 2010.

5 Surface Articulation

The problem of surface articulation unfolds in a dual fashion, on the one side articulation can be read as an ornamentation of the surface in contrast to bland, even conditions. On the other side the articulation performs as an enhancement of the structural qualities of the surface by behaving analogue to a corrugated surface.

One of the best known examples of the performance of evenly ondulating surfaces is corrugated fiberboard, the omnipresent material in shipping and handling of goods, which summarizes the qualities of this material organization: Lightweight whilst rigid in nature. The fluted and pleated conditions of corrugation are also included in the design process of the exhibition designs *The Austrian Winery Boom* and *Housing in Vienna*. Of course it is not possible to discuss the issues of corrugated surfaces in computational design without paying respect to Greg Lynns work. As early as 1999 with the Predator Project, Greg Lynn explored the sensibilities of cord du roy like surfaces. The main process used to achieve control over the fluted surface is to use the underlying geometric condition of the computational model. The direction of the isoparms, of the *NURBS* model, provides the desired control over the appearance of the surface. *Isoparms* are understood as the curves that define the surface topology of a *NURBS* object. These *NURBS* "dividers" create the surface as a sheet, with U and V directions that make up the sheet.[5] *NURBS, Non Uniform Rational B-Splines* are defined as mathematical representations for smooth curves and surfaces.[6] The combination of the information inherent in the computational model with computer controlled output techniques like CNC milling uncovers the opportunities to create a machinic ornament that oscillates between performative aspects, such as an altered structural qualities and gestures of ornamentation, opulence and voluptuous visual impressions. Once more the question of detail embedded in the surface condition comes to mind. In this case, the close relationship of computational model, fabrication strategy and the application of detail embedded in the surface geometry form a conceptual alliance resulting in a close knitted protocol of behavior reaching from the programmatic to the dramatic qualities of the surface.

6 The Austrian Pavilion at Shanghai Expo 2010

The surface and the detail of the Austrian pavilion form an intertwined condition of inter-dependent elements of architectural expression, a twofold composite of geometric rules. As with its precursors these rules are comprised of Topology and Tessellations. To briefly touch the fields of topology and tessellations the authors will rely on the definition of topology by Eric W. Weisstein: Topology is the mathematical study of the properties that are preserved through deformations, twisting, and stretching of objects. Tearing, however, is not allowed. A circle is topologically equivalent to an ellipse, into which it can be deformed by stretching, and a sphere is equivalent to an ellipsoid. Similarly, the set of all possible positions of the hour hand of a clock is topologically equivalent to a circle (i.e., a one-dimensional closed curve with no intersections that can be embedded in two-dimensional space), the set of all possible positions of the hour and minute hands taken together is topologically equivalent to the surface of a torus (i.e., a two-dimensional a surface that can be embedded in three-dimensional space), and the set of all possible positions of the hour, minute, and second hands taken together are topologically equivalent to a three-dimensional object.[7] A tiling of *regular polygons* (in two dimensions), *polyhedra* (three dimensions), or *polytopes* (n dimensions) is called a tessellation.[8] As described in the previous section these conditions of geometry where applied in the design of agglomerated components for exhibition designs. Both of the described conditions, topology and tessellation, where applied in the design of the Austrian Pavilion for the Shanghai Expo 2010.The topological body of the Austrian Pavilion was the result of a rigorous evolutionary process based on programmatic as well as issues of sensibility. To create this evolution the software TopMod was used. TopMod is a topological mesh modeling software developed primarily by Ergun Akleman, which possesses a Python language scripting interface which enables the designer to control procedures by the application of scripts. The resulting evolution had exactly one hundred cycles, resulting in a topological body that provided the desired spatial qualities: seamless circulation, recessed entrance, recessed roof-garden, autonomous entrance to the restaurant. The candidates for further development where all additionally examined in Ecotect in order to understand their performance under eco-logical pressure like Solar radiation, wind pressure and especially the acoustic qualities of the central space. In order to rigorously follow the design principles of the project it was necessary to scrutinize the topological surface, in order understand how pressures of structure as well as conditions of illumination can be seamlessly integrate into the surface condition. The authors would like to highlight two examples from the interior of the building before moving to the outside surface of the building. One of the main questions in the development of the project was how to include recessed lamps into the surface. It would have violated the entire design intention just to perforate holes into the ceiling to mount down lights in the restaurant area, so a topologically correct solution had to be found. In order to achieve a

Fig. 4 Restaurant of the Austrian Pavilion: teething of floor and ceiling ©SPAN 2010

seamless effect the main solution was to deform the area housing the lamp in order to create stomata. These deformed areas did not only provide the necessary volume to house the lamps, they simultaneously served as control field for the cone of illumination of the lights. In terms of fabrication the deformed areas where CNC milled out of foam and included in the surface before the finishing paint was applied, thus creating a composite surface made of extruded polystyrene and gypsum board. A second example about the performance of the surface is the way how ceiling and floor are connected with each other forming a teething condition. The computational model of the interior was deformed in these areas to achieve two specific conditions: on the one side structural stresses made it necessary to introduce support for the steel construction at these areas, additionally the teething condition provided a spatial subdivision, structuring the area of the restaurant into specific areas. Teething can be defined as the connection of disparate surfaces by co-planarity or blushing across surfaces. The term *teeth*, describes any connection between surfaces where two curves are tangent or have coincident control vertices. These connections allow movement, either literal or visual, across surfaces in a similar way that two level floors can be aligned in a co-planar fashion. At a smaller detail scale, teeth are a method of connecting disparate surfaces using their inflections in a similar manner to where interlocking joints are made through dado, mortise and tenon, or box cuts and interlocking joints. This technique of inflecting two surfaces together as a method for connecting them along coincident surfaces owes a great debt to Harry Cobb's theory of folding creases across disparate materials as a way of connecting materials into a monolithic curtain wall.[9]

To resume the conversation on the design of the Austrian Pavilion it is necessary to highlight that in the case of the pavilion the scale of the varying interior and exterior elements was dramatically altered. In terms of volume and mass of the construction, as well as in the size of the tiles of the tessellation. This was done in order to achieve a quasi seamless appearance of the surface. In fact the contrary is the case. More than 10 Million tiles amass into 60 Million joints, which due to its minuscule size lure the eye to believe from a distance that the surface is seamless. The advantages of this approach can be described in a trefoil way: in terms of architecture design, in terms of construction method and in terms of economic revenue. The architectural benefit lies in the ability to create an elegant surface which connects the interior and exterior in a continuous fashion. The topological surface allows for a continuous circulation from the outside to the interior and back again. A major obstacle, which was still not solved with the design of the pavilion, is the gradient transition between interior and exterior. Two questions arise out of that problem: Number one being the question about the transition of materiality between exterior and interior, from porcelain to gypsum board, if we take the materials used in the pavilion. The other question is about where to put the thermal barrier in the apertures of the building. In the case of the pavilion the problem was solved by positioning the glazing in the turning points of the topologic geometry of the pavilion. This can best observed at the entrance glazing of the pavilion. In fact the glass responds to the geometry like a soap-film would fill the hole in a physical model, analogue to the behavior of material computation. The glass was executed in a frameless fashion in order to enhance the film-like nature of the thermal barrier. The transition in materiality can be criticized in retrospect as the barrier forms a clear cut from the porcelain of the façade to the finishing of the interior spaces in gypsum board and paint. It is only fair to speculate about the opportunities to solve such a transition in future projects, especially with the use of parametric tools, as already included in the design and execution phase of the pavilion, the opportunities to address the problem of transition conditions in architectural space are multiplicious. In terms of construction method the tessellation of the surface helped to achieve the desired surface condition without compromising. It was not necessary to alter, or change the design in order to make it "buildable". Au contraire, aspects of advanced construction methods for the building where included from the very beginning of the competition into the design philosophy of the pavilion. The authors in general are very interested in advanced construction methods emerging from fields of expertise outside the architectural realm, or outside the general conventions of the building industry. However, it is also possible, and the experience of building the pavilion proved that, to use conventional construction methods in unusual ways to achieve novel architectural effects, resulting in affect conditions of the space. The combination of parametric modeling tools, with a very open construction company (in the case of the Austrian Pavilion: AMCC Alpine Meyereder.) resulted in an efficient and straight forward construction phase of the building. To illustrate the use of parametric tools in the building process the authors will rely on the tiled porcelain façade as an example. The façade reflects the allover concept of the design about topology, continuity and transitions by the use of a colored gradient. The gradient changes from red to white and to red again. The material used, porcelain, was a conscious decision by the architects to reflect the relation between China and Austria in an historic dimension. A very little known fact is that Austria, and specifically Vienna, was the second place in Europe with an own Porcelain Manufactory. The renowned Du Paquier factory established 1718 was also famous for being the first Porcelain Manufacturer to introduce the color red in the glazing. The precious material is praised in both cultures, the Chinese and the Austrian, and serves as a perfect bond between the cultures. In a series of Grasshopper definitions the size of the tiles was determined, as it was necessary to find the right balance between the ability to cover double curved surfaces without producing folds or inconsistencies as well as the simple handling of the mats with the tiles on top. The hexagonal shape of the individual tiles was selected because of its ability to avoid the presence of specific directions in the tiling of the surface, as well as for its flexibility in the connection of joints. The entire surface was divided into patches of thirty by thirty centimeters, the module of the tiling mats. The mats themselves where divided into five different percentages of color, from 100% white to 100% red. This was the basic for the gradient between red and white. Again using the parametric tool Grasshopper, the authors defined the position of the different mats to achieve the desired transition effect on the surface of the building.

The continuous covering of the surface composite with one uniform material has a variety of advantages. In terms of construction it reduces the possibilities to penetrate the layer that repels the intrusion of humidity underneath the uppermost surface. The topological quality of the architectural body reduces the possibility to break the water repelling barrier, as the soft curves facilitate the application of foils. In contrast to the prevailing convention of rectilinear surfaces where sharp edges and corners proof over and over again to be very sensible to water breaches. Conceptually speaking the subdivision of the topological surface of the Austrian Pavilion can be described as a whole to part relationship idea whereas the architectural body is subdivided into individual panels. The concept of whole to part relationships and the subdivision into individual panels of course brings along ideas about the panelization of topological bodies, which is also described as *Topo-Panelization*. Topo-Panelization is the development of a language of surface based on architectural concerns of assembly rather than sculptural interests on the monolith. It invests smooth surfaces with varied articulation, potential systems of aperture and an integral relationship to structure.[10] In fact panelization and the intimate relationship of body, panel and joint echoes the line of thought formulated by Gottfried Semper in his book *Der Stil in den technischen und tektonischen Künsten oder Praktische Ästhetik* where he discusses the nature of the seam. The unassuming nature of this heading should not cause anyone to underestimate its significance in relation to art and style. The seam is an expedient that was invented to join pieces of a homogeneous nature –namely surfaces- into a whole. Originally used in clothing and coverings, it has through an ancient association of ideas and even through linguistic usage become the *universal analogy* and symbol for any *joining of originally discrete surfaces* in a tight connection. A most important and prime axiom for artistic practice is most simply, most originally, and at the same time most cogently expressed in the seam –*the principle of making a virtue out of necessity*. It teaches us that anything that is must be patchwork, because the material and the means at our disposal are insufficient and should not be made to appear otherwise. If something is originally separate we should characterize it not as *one* and *undivided* but, by deliberately stressing how the parts are connected and interlaced towards a common end, all the more eloquently as coordinated and unified.[11]

7 Resolution

Resolution in this debate is defined as the process or capability of making individual components of an object distinguishable. This differentiation is dependent of the scale and observation distance, rendering a body comprised of components into a surface, in accordance to the relation of the size of its parts to the entire figuration. To illustrate this we can rely on two examples, the coarse resolution of components present in an example such as the Paxton Crystal palace, and the high resolution of components in a project such as the court roof of the British Museum in London by Foster & Partners. These two examples express the two differentiating approaches for the solution of a problem following two distinctive paradigms of material aggregation. On the one side we can consider the Paxton Crystal Palace to be the pinnacle of the 19th century cast iron constructions with its large scale, industrialized framework and glass construction. The second example, the glass roof, covering the court of the British Museum in London, marks the initiation of the successful use of parametric modeling techniques, where the notion of

Performative Surfaces 235

identically repeating, discrete modules is abandoned in favor of subtly varying, continuously changing components that react to specific environmental pressures within a lattice of higher resolution. In the case of Fosters roof, these parameters included the transition between a circle and a square as well as the customer's desires: a maximum of visibility of the sky, construction period while the museum

Fig. 5 South Façade, Austrian Pavilion, Shanghai Expo 2010, © SPAN & Zeytinoglu, photo © Matias del Campo

keeps open, and the inability to touch the structural system of the 19th century buildings forming the court.[12] Considering these examples we can discuss the gradient transition of components to surface in architectural environments. Especially with the emergence, and popularity of parametric tools in design the subject of surface populations became a ubiquitous technique, which can be related to concepts of nonlinear conditions. To recapitulate the aspects of the problem of the nonlinear one can rely on ideas of intensive flows of energy in and out of a system. Pushing this energy flows *far from equilibrium*[13] results in a manifold of possible results. The consequence being a multiplicity of coexistent forms in varying complexities based on statics, periodic and chaotic attractors instead of reduced forms based on simple forms of stability. Additionally it can be stated that as soon as a system transforms from one stable condition to another, forming a *bifurcation* in the process, minuscule fluctuations in the flow of energy can influence the result extensively. A system whose dynamics are *far from equilibrium* can be described as a nonlinear system, bearing features such as attractors and bifurcations and enhancing the strong mutual interactions (or feedback) between components[14].

Part to Whole Relationship, Precursors

In order to discuss a contemporary notion in the problem of panelization, whole to part relationships, joints and resolution the authors will rely on two examples: The Predator Project by Greg Lynn and the Hungerburgbahn project by Zaha Hadid Architects. Greg Lynn´s Predator was commissioned by Jeffrey Kippnis as an exhibition piece for the Wexner Center in Columbus Ohio. The Predator, which Greg Lynn conceived together with New York based artist Fabian Marcaccio forms an entity oscillating between spatial conditions and chromatic effects, *a voluptuous painting/architecture mutant hybrid*.[15] The entire body is comprised of 35 rings subdivided into 250 individual panels. The panels where covered with foil bearing the painting by Marcaccio and subsequently thermoformed into curved panels. Apart from the surface articulation achieved by computationally modeling the surface, the milling artifacts produced a highly articulated, corrugated surface. The figuration of the installation describes roughly an S-shaped ondulating tube with varying diameter. The vertical striation of the body gradually transforms into a louvered, shredded surface allowing to enter the installation and to experience the luminous, colorful translucent effects generated by the combination of curvilinear gestures and opulent coloration. In opposition to the before mentioned example of the Paxton Chrystal palace where a regime of repetition forms the rhythm of the structure, the Predator follows a logic of continuous differentiation, where the whole represents the main order, and the parts form individual components of the whole. In the light of computational design techniques generating the performance of details in the surface, the Predator Installation, with its corrugated surface condition, emerging in the fabrication process, can be considered one of the crucial examples.

In order to describe the difference between procedures of tectonics applied to curvilinear gestures, and the concept of continuous formation applied to the Austrian Pavilion, the example of Zaha Hadid´s Hungerburg Bahn in Innsbruck will serve as an example. The project consists of four stations that transport people from the congress center in Innsbruck to the Hungerburg. The continuous, fluent language of the stations, behave in a delicate interplay between the mountainous surroundings. Zaha Hadid Architect´s contribution to the discourse gravitating around the inception and execution of advanced concepts of architectural design cannot be respected enough, in the case of the Hungerburgbahn however, the concept follows a lineage of subdivision of surface leaning towards prevailing methods of structural glass systems applied in planar conditions. The achievement to clad these curvilinear bodies with glass can, and should be respected for addressing the problem of surface enclosures of double curved bodies. The subdivision of the body in panels reflect in way the appearance of isoparms in the computational model, transmutating the model into a specific sensibility of the constructed reality. Exactly this moment of transmutation, however addresses the problem of detail in double curved conditions. In opposition to a planer structural glazing, the continuous change in the surface condition of the Hungerburgbahn stations asked for differentiating points of attachment for the glass-panels. Obviously the subdivision of the surface was a conscious decision engaging in a sensibility of joints as ornament that also supports a specific articulation of the architectural body. The alternative modes to the described construction method (which resonates techniques of construction of the industrial age) would be to work with laminating and gluing to achieve a seamless continuous surface or to reduce the size of the panels till they start to lose their ability to form sightlines along the joints guiding the eyes in specific directions.

8 Conclusion

In the light of contemporary development in computational form finding processes, wither they be by the application of naïve algorithms or through the use of sophisticated generative computational models, the question of the subdivision of space into construction elements remain. This paper focuses on the relation between surface condition and detail in order to speculate upon the opportunities to fold the program of detail into the surface condition. The built example of the Austrian Pavilion at Shanghai Expo serves as proof of concept that by reducing the mass of details on the façade a multitude of prevailing problems in architectural production were addressed, such as the shift from horizontal to vertical in the materiality, the problem of resolution, the problem of gradually changing chromatic effects, the transition of materiality between interior and exterior and the question about fenestration in complex curved geometries. The combination of parametric modeling tools and the linear communication via 3D models facilitated the dialogue with the construction site. The idea, to reduce detail as additional element of the design, and to embed the performance in the geometry of the surface, proofed to be a valuable concept, which not only provided for an uncompromised, rigorous result from the design phase to the executed building but also helped to reduce construction time and costs. In a larger scale the concept of detail performance in the surface behavior means a paradigmatic shift in the conception of architectural space, from a tectonic, additive system, towards a lineage of forming.

References

1. Kipp, A.: Technik im Barock, http://www.museumsprojekte.de/Start/Service/Skripte/biblio.htm (visited June 8, 2010)
2. Hersey, G.: Architecture and Geometry in the Age of the Baroque, p. 5. The University of Chicago Press, Chicago (2001)
3. de landa, M.: Uniformity and Variability an Essay in the Philosophy of Matter, http://evans-experientialism.freewebspace.com/de_landa01.htm (visited June 10, 2011)
4. de Landa, M.: Intensive Science and Virtual Philosophy, Continuum, New York, p. 12 (2002)
5. Lammers, J., Gooding, L.: Maya 4.5 Fundamentals, p. 148. New Riders, Indianapolis (2003)
6. Gould, D.A.D.: Complete Maya Programming, An extensive Guide to MEL and the C++ API, p. 476. Morgan Kaufmann Publishers an imprint of Elsevier, San Francisco (2003)
7. Weissstein, E.W.: Topology. From MathWorld–A Wolfram Web Resource, http://mathworld.wolfram.com/Topology.html
8. Weisstein, E.W.: Tessellation.From MathWorld –A Wolfram Web Resource, http://mathworld.wolfram.com/Tessellation.html
9. Lynn, G., Lynn, G.,, H.: Architectural Laboratories, p. 25. NAi Publishers, Rotterdam (2002)
10. Balliet, C., Buck, B.: Visual Catalog: Greg Lynns Studio at the University of Applied Arts Vienna, p. 14. Springer, New York (2010)
11. Semper, G.: Der Stil in den technischen und tektonischen Künsten oder Praktische Ästhetik, Verlag für Kunst und Wissenschaft, Frankfurt, S77–S79 (1860)
12. Cook, M.: Digital Tectonics, Historical Perspective – Future Prospects. In: Leach, N., et al. (eds.) Digital Tectonics, pp. 41–49. Wiley-Academy, Chichaster (2004)
13. de Landa, M.: A Thousand Years of Nonlinear History, p. 14. MIT Press, Cambridge (1997)
14. del Campo, M., Manninger, S.: In: Goldenberg, E. (ed.) Rhythm & Resolution, from the upcoming book Pulsation. J. Ross Publishing, Fort Lauderdale (2011)
15. Beal Center for Arts & technology webpage (June 12, 2011), http://beallcenter.uci.edu/exhibitions/predator.php

ICD/ITKE Research Pavilion: A Case Study of Multi-disciplinary Collaborative Computational Design

Moritz Fleischmann and Achim Menges

Abstract. Material has the ability to *compute*. A material construct can be considered as the equilibrium state of an intricate network of internal and external forces and constraints. In architecture, this understanding is of particular relevance, if the elastic bending behaviour of material is allowed to play an active role in the design and construction process. As form can no longer be merely geometrically defined and described separately from structural behaviour that governs its material gestalt, the role of the designer in the process of computer-aided architectural design based on such form-finding principles changes considerably.

Based on a realized project, the *ICD/ITKE Research pavilion*, the paper will present insight to alternative computer-based design and information modelling strategies for such material systems: The pavilion was designed, planned and built as a multi-disciplinary project with a strong focus on information modelling and feedback between design, structural simulation and computer-aided manufacturing.

1 Integral Computational Design

In todays practice, computational tools are still mainly employed to create design schemes through a range of design criteria that leave the inherent characteristics of the employed material systems largely unconsidered. Ways of materialisation, production and construction are strategized only after a form has been defined, leading to top-down engineered, material solutions. Based on an understanding of form, material and structure, not as separate elements, but rather as complex interrelations that are embedded in and explored through integral generative processes, the research presented in this paper demonstrates an alternative approach to computational design [1]: here, the computational generation of form is directly driven and informed by physical behaviour and material characteristics as well as by the

---

Moritz Fleischmann · Achim Menges
Institute for computational Design (ICD), Stuttgart University, Germany

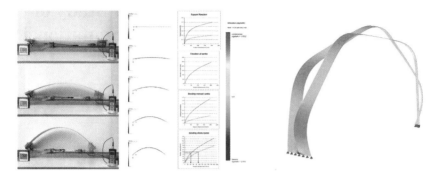

Fig. 1 Physical / digital test of elastic plywood bending behaviour (left) & Simulation of system behaviour (right)

fabrication constraints [2]. The structure is entirely based on the elastic bending behaviour of birch plywood strips (Fig. 1). These strips are robotically manufactured as planar elements, and subsequently connected so that elastically bent and tensioned regions alternate along their length. The force that is locally stored in each bent region of the strip, and maintained by the corresponding tensioned region of the neighbouring strip, greatly increases the structural stiffness of the self-equilibrating system. This is explained by the fact that the bent arch induces tensile stress into the straight arch. This tension pre-stress increases the lateral stiffness of the straight arch and hence the geometrical stiffness of the entire system.

2 Prototype Structure

The spatial articulation and structural system is based on a half-torus shape. Defining the urban edge of a large public square, it touches the ground topography that provides seating opportunities on the street facing corner (Fig. 2). In contrast to this, the torus side which faces the public square is lifted from the ground to form a

Fig. 2 Exterior (left) & Interior view (right) of prototype structure

free-spanning opening. Inside, the toroidal space can never be perceived in its entirety, leading to a surprising spatial depth that is further enhanced by the sequence of direct and indirect illumination resulting from the convex and concave undulations of the envelope, which derives its form from the equilibrium state of the embedded forces (Fig. 3). The combination of both the stored energy resulting from the elastic bending during the construction process as well as the morphological differentiation of the joint locations, allows a very light, yet structurally sound construction. In this lightweight structure bending is not avoided but instrumentalized, here referred to as bending-active structure. The entire structure, with a diameter of over 10 meters, can be constructed using only 6.5 mm thin birch plywood sheets with a total construction weight of only 400 kg.

The half-torus timber shell was constructed out of 80 individual strips (Fig. 3). Each of the strips consisted of 6 or 7 segments and was coupled with its neighbouring strip to form a self-stabilizing arch (Fig. 1). The coupling occurred at either end of the segments along the side of each strip. The unequal lengths of the neighbouring coupled segments forced the longer segment into the post buckling shape of an elastica arch. The final torus shape was then constructed by coupling 40 self-stabilizing arches in a radial arrangement. In order to prevent undesirable local stress concentrations as well as the adjacency of weak spots between neighbouring strips, the coupling points need to shift their locations along the structure, resulting in 80 different strip patterns and resultantly 500 different segments.

3 Information Model

Because of the very short timeframe available to develop and build the Pavilion and its unique characteristics in terms of material behaviour, the team decided to integrate as many aspects of the design process as possible into a central information model. The university's recently acquired robotic manufacturing workshop allowed fabrication of most of the elements needed for construction on site. This opened the possibility for a direct transition from the digital domain to the physical artefact.

Fig. 3 Robotic manufacturing of plywood strip (left) & Assembly on site (right)

Therefore, as opposed to conventional CAD-modelling strategies in the building industry we developed a unique computational information model, which consisted of 2 integral parts: A parametric model[1] which integrated the knowledge derived from physical form-finding experiments into a geometric output in form of a polyline skeleton and secondly a multi-subroutine script[2] which used the parametric models polylines as an input to create the geometric model[3]. This was achieved by continually developing the information model parallel to the design process, informing it through early physical form-finding experiments, embedding appropriate methods for geometric description of physical behaviour and successively expanding its functionality as the project progressed. Subsequently,, individual subroutines existed, within the information model, for all aspects of the pavilion design: From creating a 3-dimensional NURBS-Surface model (*Geometric model*) for visualisation & solar analysis to generating unrolled 2D strip geometries for FEM-simulation and robotic fabrication (Fig. 3).

This aspect of the project presents a novel approach to conventional Building Information Modelling (BIM). By developing highly specialised, individual subroutines, it was possible to derive the data necessary for the design, fabrication, construction & assembly of the structure without needing to revert to preconfigured building blocks / elements as it is usually seen in commercial BIM software. For the planning of the prototype pavilion the integral computer-based design strategy was one of the essential measures which helped to coordinate and mediate various design and fabrication factors. This information model combines relevant material behavioural features in form of abstracted geometric relationships that were transferred into parametric principles (Fig. 8).

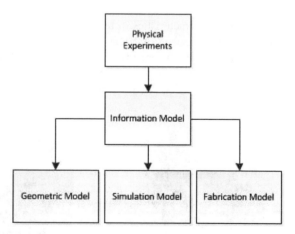

Fig. 4 Conceptual Information Modelling Scheme used for the Pavilion

---

[1] Developed in Grasshopper, a parametric extension to McNeel's Rhinoceros CAD Software;
http://www.grasshopper3d.com/
[2] Written in McNeel's RhinoScript Scripting language.
[3] Nurbs Surface Geometry.

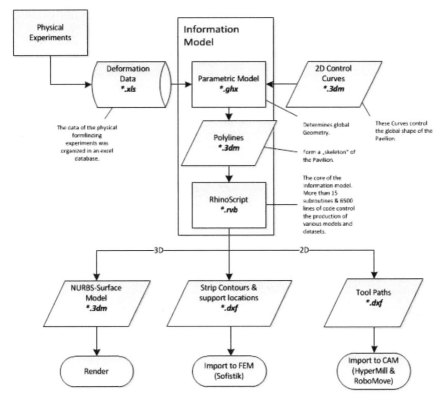

Fig. 5 Detailed Information Modelling Scheme used for the Pavilion

3.1 Representing Material Behaviour by Geometric Means

In order to capture the buckling behaviour of the thin plate material under compressive load, a Curve-Matching Algorithm was develpoed. It formed a central aspect of the *Information Model* and was based on data derived from physical form-finding experiments. The goal of the form-finding experiments was to measure the increase and limits of the rise of an arch that is observed in a single span beam under compression in its post buckled shape (Fig. 1).

In a first step, a curve-matching algorithm was developed in order to approximate these shapes computationally. The aim was to find a geometric description of the longitudinal system-axis that was controlled by a small set of parameters: Subsequently, a 2-dimensional BSPLINE-Curve was constructed with 3 control points. The first A and last control point C defined the start and endpoint of the curve, whereas the 2nd control point B was used to create an arch-like shape. This was achieved by placing B at midpoint between A & C at curve parameter

$$t = 0.5 \tag{1}$$

Fig. 6 Variation of local parameters cause changes in overall geometry

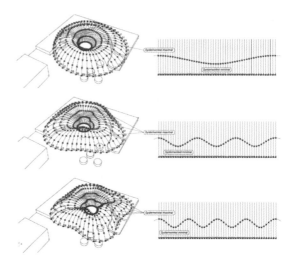

Fig. 7 Curve-Matching Algorithm

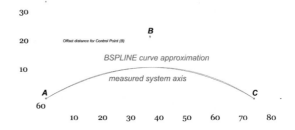

Fig. 8 Parametric description of a series of timber strips. Local buckling condition determines overall geometry.

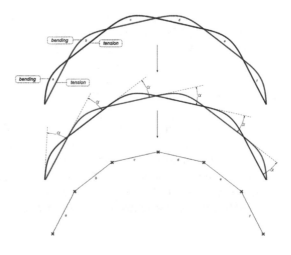

along the curve *AB* and stepwise offsetting it perpendicularly in one direction. The offset amount (length) as well as the curve weight associated at *B* were enough to sufficiently describe the desired longitudinal system-axis computationally (Fig. 7). As the inherent logics of computational BSPLINE-curves are derived from mechanical spline behaviour it could be expected that matching the results from physical experiments lead to a sufficient correlation of the mechanical behaviour and geometric data embedded in the design model.

3.2 Collaborative Design and Information modelling

Throughout the entire design process, the functionality of the Information model was expanded successively. Each phase (design, detailing, fabrication & construction) resulted in a functional expansion of the information model. Internal feedback ensured that the increasing amount of constraints were considered while design intentions were maintained.

The final design of the pavilion is the product of this process, which started with the seemingly simple task of embedding the geometric logics of the buckling behaviour of thin plate timber laminae under compressive external loads into a script.

Even though modelling such structural behaviour could have been realized through the use of simulation software, the team decided to write a geometric curve-matching algorithm in Rhino based on the physical formfinding experiments from scratch, rather than to simulate each iteration of the design during the decision making process in FEM-Software. The reasons were not only the immediate feedback, but also the expandability of the information model: Once incorporated into the information model, the data could be used to create fabrication layouts asf., without the need to im- or export data. This decision was only possible because of the simplicity of the system: As each curved strip was only bent in one direction (and not twisting along its longitudinal axis), the information model could be abstracted to series of lines (for planar strip segments) & B-Spline Curves (for the strip segments, which were bent).

The trade-Off of this approach was the lack of structural feedback beyond the strip segment: The effects of deformation under self-weight for example, were not embedded.

Yet, this proved not be a disadvantage: Comparing the geometric output of the information model to the FEA simulation showed only minor deviations [4]. The link to the FEA Software was written as a separate subroutine inside the information model in order to perform structural analysis for the building permission and strip dimensioning.

At the time of the design, physics-based CAD environments or Plugins such as Kangaroo[5] or Vasari[6] did not exist yet. Furthermore do most other physics-based

---

[4] Paper submitted acceptance for IASS 2011 by Julian Lienhard, Simon Schleicher and Prof. Jan Knippers, ITKE.

[5] A Plugin for Grasshopper written by Daniel Piker, http://kangaroophysics.com

[6] A LABS Product of the AutoDesk Group with "Nucleus" technology developed by Jos Stam; http://labs.autodesk.com/utilities/vasari

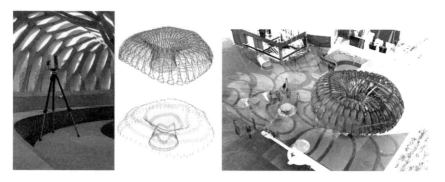

Fig. 9 Full scale scanning and geodesic measurement (left) & Pointscan Model in Rhino (right)

modeling engines lacked the ability to model bending behaviour. Writing such an engine from scratch was not feasible given the timeframe for the construction of the pavilion. The question whether embedding physics-based structural feedback into the information model might be an alternative approach to the information modeling strategy provided here, are currently being investigated[7].

> Because of the team-internal expertize and the collaborative environment, the concept to create an information model that could be used to output a geometric design model or create the input for a separate structural analysis was very beneficial as the feedback between team members was almost immediate. Whenever the deviation between the simulation model and the Design model became too large, changes were made within the information model in order account for these.

4 Research Verification

From a design process point of view, the approach to gather relevant information in a centralized model proved to be very successful. But beyond the short feedback cycles between the team members gained through this, we were furthermore interested in how geometrically similar each of the produced models was to the actual built artefact.

The development and construction of the research project allowed verification of the presented computational design and fabrication process by comparing the following three models which were all derived from a centralized information model:

---

[7] At the Institute for computational Design (ICD), Prof. Achim Menges (PHD Candidate: Moritz Fleischmann).

- the computational *design model*.
- the related FEM *simulation model* derived by structural engineers.
- the actual geometry based on a *pointscan model* of the constructed pavilion measured by geodesic engineers.

4.1 Comparison between Geometric Model and Pointscan Model

The Geometric model did not embed behavioural features beyond the description of each locally bent segment. It intentionally factored out additional parameters such as the weight of the overall structure, relaxation of wood due to exposure to humidity & sunlight as well as the effects of additional deformation of the self equilibrating arches when coupled to neighbouring arches. All of these aspects do effect the overall geometry. The observed deviation between pointscan and geometric model is therefore tolerable and was expected to a certain degree (Fig. 10).

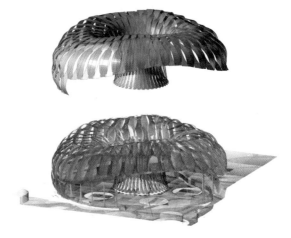

Fig. 10 Comparison of Design Model (Top) & Pointcloud Model (Bottom)

4.2 Comparison between Simulation Model and Pointscan Model

Even though the previously mentioned factors were embedded in the simulation model, a marginal deviation to the data of the Pointscan was observed. The minor discrepancies are due to the (necessary) simplification of the model: e.g. Mechanical behavior on the detail level, such as the slip at each strip-to-strip connection, was not accounted for. Furthermore, the geometry of the individual strips were abstracted, as the curved outlines proved to be a challenge for the necessary meshing that proceeds every FE-Simulation.

Acknowledgements. The Research Pavilion was a collaborative project of the Institute for Computational Design[8] (ICD, Prof. Achim Menges) and the Institute of Building Structures

---

[8] ICD website: http://www.icd.uni-stuttgart.de

and Structural Design[9] (ITKE, Prof. Jan Knippers) at Stuttgart University, made possible by the generous support of a number of sponsors including: OCHS GmbH, KUKA Roboter GmbH, Leitz GmbH & Co. KG, A. WöLM BAU GmbH, ESCAD Systemtechnik GmbH and the Ministerium für Ländlichen Raum, Ernährung und Verbraucherschutz Landesbetrieb Forst Baden-Württemberg (ForstBW). The project team included Andreas Eisenhardt, Manuel Vollrath, Kristine Wächter & Thomas Irowetz, Oliver David Krieg, dmir Mahmutovic, Peter Meschendörfer, Leopold Möhler, Michael Pelzer and Konrad Zerbe. Responsible for the scientific development were Moritz Fleischmann (project management), Simon Schleicher (project management), Christopher Robeller (detailing / construction management), Julian Lienhard (structural design), Diana DSouza (structural design), Karola Dierichs (documentation).

References

1. Menges, A.: Performative Wood: Integral Computational Design for Timber Construction. In: Proceedings of the 29th Conference of the Association For Computer Aided Design In Architecture (ACADIA), Reform: Building a Better Tomorrow, Chicago, USA, pp. 66–74 (2009)
2. Menges, A.: Material Information: Integrating Material Characteristics and Behavior in Computational Design for Performative Wood Construction. In: Formation, Proceedings of the 30th Conference of the Association For Computer Aided Design In Architecture (ACADIA), New York City, USA, pp. 151–158 (2010)

---

[9] ITKE website: http://www.itke.uni-stuttgart.de

Metropol Parasol - Digital Timber Design

Jan-Peter Koppitz, Gregory Quinn, Volker Schmid, and Anja Thurik

1 A Shading Structure for Seville

The redevelopment of the Plaza de la Encarnación and the design of the unique free formed timber mega structure "Metropol Parasol" (Fig. 1.1) started with a

Fig. 1.1 Plaza Mayor in Seville with the Metropol Parasol

Jan-Peter Koppit · Gregory Quinn · Volker Schmid
Arup

Anja Thurik
FinnforestMerk

design competition in 2004. The aim was to provide a new underground museum for the newly excavated roman mosaics, space for shops and the original market tands at ground floor level and a public plaza at 5 m above ground, including bars and a restaurant. The Berlin architect Jürgen Mayer H. together with the engineers of Arup submitted the winning scheme. Most visible is the huge timber structure resembling six merging mushrooms of up to 28 m in height and 150 m in total length, which provide the shading for the new centre point of urban live. Meandering walkways on top of the parasols and a restaurant at 21,5 m above ground invite tourists to enjoy the view on the old town.

2 Shaping the Parasols

The geometry for the structure is based on a free-form, outlining the tree-shaped shading structures. The individual laminated wood plates (LVL) are generated by cutting vertically in an orthogonal 1.5x1.5 meter pattern through the free-form.

Fig. 2.1 Outline of the Parasols

Fig. 2.2 Rhino-Model of the wooden structure for the Metropol Parasol

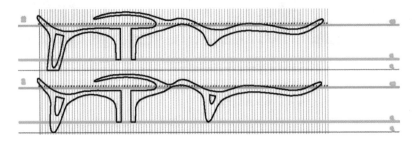

Fig. 2.3 Section profiles for a selection of the wooden elements

3 The Hybrid Structure

Different materials are employed in the structure depending on the various architectural and structural demands. The foundation and the cylindrical elevator shafts below the platform restaurant are made of concrete. The museum area was spanned with composite trusses of steel and reinforced concrete held together beneath by tie rods. A composite steel structure bears the weight of the restaurant and is supported by various slanted struts made of hollow steel sections, which follow the path of the outside stairs and are connected to two concrete cores.

The timber plates are 1.5 to 16.5 meters long; the width of the Kerto-Q LVL-plates varies from 68 to 311 millimeters. The parasol elements reach a maximum height of about 3 meters, while the largest construction piece, in the "trunk," measures 16.5 x 3.5 x 0.14 meters. Overall, there are about 3,400 elements, with a total volume of about 3,500 cubic meters of laminated veneer lumber.

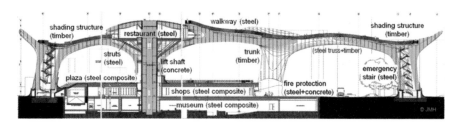

Fig. 3.1 Cross section of the hybrid structure for the Metropol Parasol

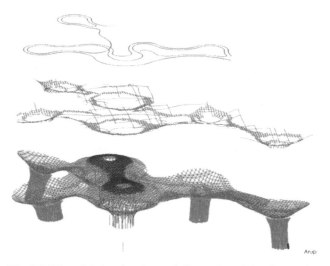

Fig. 3.2 FE-model showing the steel diagonals and the observation walkways

Since the shading structure does not have a closed, stiffening roof, but rather one that is open at the top and exposed to the elements, additional steel diagonals were necessary to stabilize the wooden structure so that it could bear weight. A clever, well-defined arrangement of the diagonals made it possible to achieve bi-directional shell action in the wooden grillage.

4 Innovative Finish with a Polyurethane Membrane

The parasols' wooden structure is at the mercy of the elements. In order to protect the wood, the laminated veneer lumber was sprayed with a two to three millimeter thick layer of 2K polyurethane (Fig. 4.1). Thanks to its outstanding flexibility and its excellent adherence to wood surfaces, this layer can help to prevent possible fissures in the wood. At the same time, the 3mm thick PUR layer is sufficiently vapor permeable. Together with the light ivory-colored topcoat of paint that serves as UV protection, the polyurethane coating lends the wood an entirely new surface quality. This new combination of PUR and wood provides the engineers and architects completely new ways to treat wooden structures.

Fig. 4.1 2-3mm thick 2K polyurethane coating sprayed onto Kerto-Q with glued-in threaded

5 The Connection Details

In accordance with the original cutting pattern of the structure, all of the joints in the ground plan are at right angles. In elevation, however, all of the 2700 joints are at different angles. The wooden joints have to bear forces of up to 1.3 MN. Because the foundations load-bearing capacity is limited, the lightest possible connecting detail is required. Since each joint is different, a flexible, modular system has to be developed. The connection has to accommodate tolerances on site and be easy to assemble. Since every connection is visible, their dimensions have to be as minimal as possible.

The answer to these requirements is connections that use bonded rods: a modern concept featuring great load-bearing capacity, but relatively lightweight. For the moment connections on the top and bottom sides of the elements, a special, standardized clevis connector was developed; it can be rotated and quickly bolted on at the construction site, as in steel construction. The two steel plates are interlocked via a saw-toothed connection and connected by pre-stressed, high-tension bolts to the flange. The high forces are transferred between the steel and wood via the threaded rods glued into the wood.

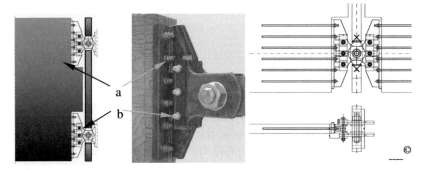

Fig. 5.1 Principle of moment connections using glued-in threaded rods (a) and pre-stressed bolts (b)

6 Heated Connections

In Seville temperatures can reach well over 40 °C in the shade. This causes problems to the epoxy resin used to glue in the threaded rods, because the resin is only approved for up to 60 °C. Arup's thermal simulations showed that temperatures of at least 60 °C or more could be reached inside the wooden structure. For this reason Arup's suggested to raise the glass transition temperature of the epoxy resin by tempering it: Working with WEVO Chemicals, Finnforest Merk developed a controlled process for heating the timber elements, including the bond lines of the rods, to about 55 °C, making it possible to increase the glass transition temperature, to well over 80 °C.

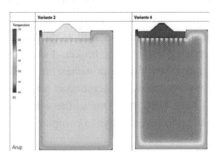

Fig. 6.1 Temperature simulation performed on 68mm thick Kerto-Q boards

7 Integrated Design and Production

7.1 Continuous Electronic Data Transfer

The Metropol Parasol project was only conceivable by means of an integrated design development among all team members; architect, structural, services and fire engineers as well as the timber contractor. A prerequisite for such a design process is the seamless transfer of electronic data between all team members including the general contractor both in Germany and Spain. Collaboration in the team begins at the very first design phase. Virtual 3D models from the architects provide geometry that is directly modified or optimised by the engineers according to analysis results.

7.2 Generating Input Values for Structural Calculations

The structural analysis of the Metropol Parasol requires highly complex three dimensional calculations. The starting point for the analysis is the architect's 3D model which provides the height and profile of the wooden elements as well as their orientation in space. With the help of a custom made computer program, the structural engineers automatically import the relevant geometric data from the model and generate an analysis model. The analysis model assigns a mass to each individual wooden element and steel connector in the system depending on the element's width. A key issue with the structural analysis of the Metropol Parasol is the sheer number of structural elements which results in a serious computational challenge, i.e. how do I obtain results if my input values are constantly changing? The final geometry of the structure depends on the forces acting at each node in the structure and these forces depend on the width and subsequent weight of the structural elements as well as the necessary connector sizes. This cycle of interdependence can only be broken if the whole structure is solved iteratively. For each iterative step the input values are extracted from the results of the previous step. Convergence of the iteration is reached when all geometric and load bearing criteria have been met. Due to the complexity of the geometry, this convergence can take up to several days to compute.

7.3 Automated Iteration Routines

The structural engineers at Arup developed a software routine in order to automate the iteration. This takes into account the thickness and weight of the wooden elements and of the connection details at each node in the structure. The thickness and subsequent weight of the wooden elements are redefined at each step of the iteration process depending on the loads in the system.

Similarly, the weight of the connectors is redefined at each step of the iteration. The various connector types which differ in weight and load-bearing capacity are defined according to the thickness, height and geometric orientation of fibres in the wooden elements with the help of a large custom-made data matrix.

Once the iteration process is complete, the results are collected and forwarded electronically to the wood contractor and the architects. This same data is then used directly for the fabrication and detail checks.

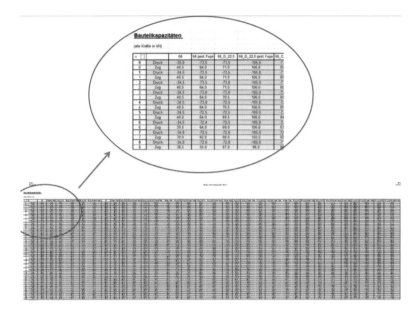

Fig. 7.1 Screenshot from the matrix that assigns connector types according to size and loading categories of the wooden elements.

Fig. 7.2 Various widths of LVL, matched to the given forces in the timber beams

After completion of detail design, the real weight of the timber structure including the connection details was compared with results from the last iteration of the FE model.

7.4 Computer Aided Manufacture

Data from the architectural model (JMH) is combined with data from the structural engineers (Arup) in order to generate a virtual object model for the timber contractor (FFM) to which further production-specific data is added. The output from this is information that can be used for remaining detail checks by specialist engineers (IB Harrer) and to produce final working drawings.

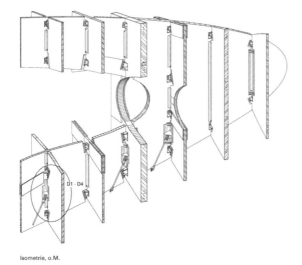

Fig. 7.3 Isometric view including connections, 3D-construction model by FFM

8 Production Sequence and Assembly

The vacuum treated raw panels, 24 and 33 millimeters thick, were bonded together into large panels, from 68 to 311 millimeters thick, in a vacuum process. A CAD team using macro programming constructed semi-automatic processes for the approximately 3,400 individual wood elements and connections. After being approved, timber elements of the same thickness were optimally nested in the slabs—taking into account the grain direction —in a process similar to cutting cookies out of dough. The elements were precisely cut down to the millimeter by a CNC-controlled trimming robot, and were milled and notched at the same time. The 35,000 longitudinal bores (65–70 cm deep) for the glued-in threaded rods were drilled manually.

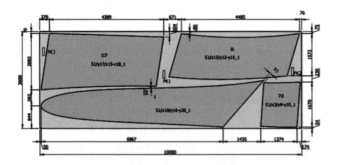

Fig. 8.1 Cutting pattern of the individual Kerto panels   © FFM

Metropol Parasol - Digital Timber Design

The fundamental idea behind the montage of the parasol structures was to keep element assembly as flexible as possible and to use simple, steel connection details to transfer the axial forces and bending moments, and also accommodate the construction tolerances. The assembly was carried out with the help of a complex scaffolding system, which was coordinated with the geometry of the parasol and its corresponding loads. In addition, assembly baskets were specially made to fit one person, matching the dimensions of the parasol grillage.

Fig. 8.2 Views of the platform and scaffolding, carpenters in custom built baskets

Performative Architectural Morphology

Finger-Joined Plate Structures Integrating Robotic Manufacturing, Biological Principles and Location-Specific Requirements

Oliver Krieg, Karola Dierichs, Steffen Reichert, Tobias Schwinn, and Achim Menges

Abstract. Performative Architectural Morphology is a notion derived from the term Functional Morphology in biology and describes the capacity of an architectural material system to adapt morphologically to specific internal constraints and external influences and forces. The paper presents a research project that investigates the possibilities and limitations of informing a robotically manufactured finger-joint system with principles derived from biological plate structures, such as sea urchins and sand dollars. Initially, the material system and robotic manufacturing advances are being introduced. Consequently, a performative catalogue is presented, that analyses both the biological system's basic principles, the respective translation into a more informed manufacturing logic and the consequent architectural implications. The paper concludes to show how this biologically informed material system serves to more specifically respond to a given building environment.

1 Introduction

Architectural material systems have long been restricted in their possible geometric differentiation due to manufacturing constraints. Primarily because of economic factors, systems with a large number of similar construction elements have been given preference. However, digital fabrication methods allow for the introduction of a high level of morphological differentiation in architectural design.

Performative morphology denotes the capacity of an artificial material system to adjust specifically to system-external and system-internal conditions through morphological differentiation (Hensel and Menges 2008). This level of differentiation is emblematic for natural systems, where a high degree of morphological variation,

---

Oliver Krieg · Karola Dierichs · Steffen Reichert · Tobias Schwinn · Achim Menges
Institute for Computational Design, University of Stuttgart, Germany

and consequently of functionality, is achieved with relatively minimal material input (Nachtigall 2004).

Thus the realization of a biomimetic design approach becomes not only practically feasible through the application of digital fabrication methods but also serves as a necessary filter, limiting possible variation in digital production through the specific information of the material system with biological principles.

The paper will present in detail a novel robotic manufacturing process for finger-joint plate structures. In the first part, architectural precedents are introduced to outline the specific manufacturing innovations that are enabled through the application off a six-axis industrial robot. Consequently the digital information chain from the parametric information model to the manufacturing of the finger-joint plate structure will be described in greater detail. The third part will focus on the information of the model with principles derived from biological plate structures and the integration of a feedback loop with FEA simulation software to verify some of the results. To conclude, both the innovative aspects of the manufacturing process itself and the specified performative aspects achieved through the biological information will be highlighted and discussed (Fig. 1.1).

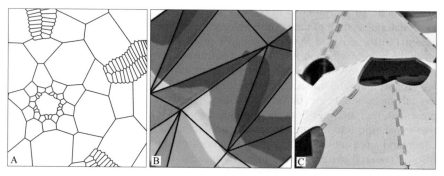

Fig. 1.1a Schematic top view of a sea urchin showing the plate arrangement **b** FEA deformation analysis of a plate structure (6.4mm thickness, spanning 140cm) under self-load (DMX = 0.057mm) **c** Robotically manufactured finger-joined plate structure

2 Design Methodology

Industrial robotic manufacturing systems show higher degrees of freedom in their range of motion than common process-specific CNC machines. The presented research project makes use of the machine's degrees of freedom and develops a three-dimensional finger-joined plate structure system (Fig. 2.1). Still, architectural systems require a high amount of various information sets, not only focusing on structural characteristics, but also considering aspects of spatial arrangements, lighting conditions, insulation, and many more. Due to this high degree of complexity, a research methodology was used that suggests informing previously defined architectural material principles with a variety of principles from biological systems into a so called bio-informed material system.

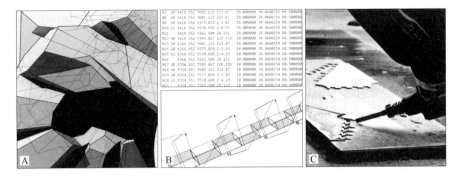

Fig. 2.1 Computational design and manufacturing process **a** Generated 3D model **b** Automatically generated NC-code and manufacturing information **c** Newly developed method of robotically manufacturing three-dimensional finger joints

Robotic manufacturing with six or more axes is once again broadening the range of manufacturing possibilities in the same way CNC machines did before, but instead of standardising joint manufacturing processes, robotic manufacturing now fully enables the use of Schindler's described Information-Tool-Technology (2007) and thus the possibilities of designing parametric joint geometries (Fig. 2.2), whose manufacturing information is directly being transferred from the generated 3D-geometry, representing the developed plate structure, by using an automated NC-code generation process.

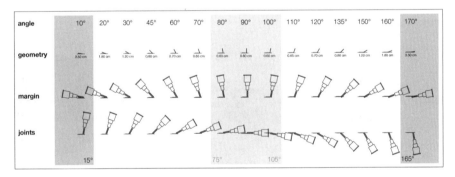

Fig. 2.2 Manufacturing constraints of the newly developed finger joint with a 20/100/20mm shank cutter. Depending on the clamping the light grey area might not apply.

3 Case Study

Finger joints connect force- and form-fitting elements through multiple interlocking teeth with a straight or tapered shape. This joint allows for connections without any additional fastener and avoids warping effects during dimensional changes (i.e. swelling or shrinking of wood) of the structural elements.

However, manual manufacturing as well as current CNC technolgy for trimming and milling machines is limited to producing finger joints for rectangular plate connections, although connections for beam structures are already highly evolved in industrial fabrication. The presented research has developed a way of rapidly yet highly precisely producing finger-joints for entirely closed plate connections at varying angles (15°-165°) by employing a 6-axis industrial robot milling at right angles with the finger joint connection and thus avoiding unnessecery gaps (Fig. 2.2). Through the close interrelation of computational form generation and robotic manufacturing, joint connection geometries not only with varying finger joint angles and miters (Figs. 3.1a and b), but also connecting three or more plates along one margin (Fig. 3.1c), are feasible and now only limited by the robot kinetics' range of motion.

Christoph Schindler's ZipShape (2007) examplarily shows the close interrelation of computational form generation and robotic manufacturing, leading to a wide range of possible geometries. However, focusing on the need for an integration of the potential applications of the introduced finger joint system into a generative computational design procedure, the question needs to be answered as to how the possible geometric differentiations can be used in an application specific manner. Biology offers a multitude of examples how local and global morphological differentiation in plate structures allow for performance-based adaptations. The wide range of computationally feasible variations of the plate-structure is thus informed and gradually specified not only through the constraints of the industrial robot, but also through embedding biological principles into the computational model.

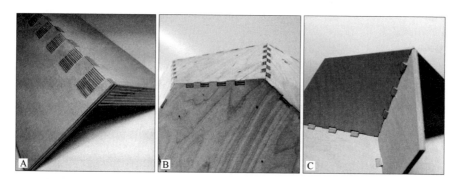

Fig. 3.1 Robotically manufactured finger joints **a** Connecting two plates at a specific angle with differentiated thickness along their margin **b** Plate structure prototype with differentiated finger joint geometries **c** Multi-dimensional connection of finger-joined plates meeting in one point.

As stated by Nachtigall (2004) the advantage of plate structures in comparison to folded structures like origami patterns lies in their topological rule of joining not more than three plates in one point. By following this principle no bending forces occur along the plate's margins, thus ensuring the system's stability. As a result, the principles of origami patterns, such as the ones researched by Buri and

Weinand (2006 and 2008), do not need to be taken into account. Instead the research on possible plate arrangements rather needs to be of a much wider range. To find a performative pattern, biomimetic principles from biological plate structures were investigated and extracted.

Plate structures can be found in cancellous bone, turtle shells, skulls, and most importantly in echinoids, such as the sea urchin and sand dollar. The taxonomic class of the *Echinoidea* was of particular interest during the research on plate structures due to the significant and distinct plate arrangements forming the sea urchin's rigid test, which is composed of fused plates of calcium carbonate covered by a thin dermis and epidermis (Barnes 1982). Not only their global plate topology, but also their global and local constructional and functional morphology were used to inform the research process. For that purpose, a performance catalogue is developed that allows the designer for the integration of architectural performance criteria and the parameterized biomimetic principles. The catalogue investigates the stated biological systems in different performance criteria, of which the most important ones are:

- Local plate arrangement (Fig. 3.2)

Biological principle of echinoids: Due to a maximum of three plates meeting in one point, echinoid's skeletal plates are mainly exposed to shear force and minor degree of bending.

Architectural translation: Following this principle the load transfer is reduced to shear force along the connection axis, for which the finger joints are particularly suitable.

- Local load transfer (Fig. 3.3)

Biological principle of echinoids and turtle shells: The plate's suture joints are forming a finger joint-like shape, while maintaining space between the opposing margins.

Architectural translation: Manufacturing-based tolerances can be useful for both an easier assembly procedure and absorbing movements.

- Global plate arrangement (Fig. 3.2)

Biological principle of echinoids: The global arrangement of the sea urchin's plate leads to a five-fold symmetry, which is achieved by only changing the plate's size.

Architectural translation: Differentiation in the plate size can be seen as a parameter reacting to structural and architectural constraints.

- Edge formulation

Biological principle of echinoids and turtle shells: Skeletal plates become smaller in order to increase the suture joint surface and thus the ability to absorb dynamic impact forces.

Architectural translation: This principle can be adapted by controlling the plate's sizes through different parameters reacting to possible impact and wind forces.

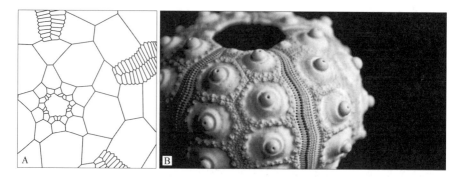

Fig. 3.2 Local and global plate arrangement **a** Schematic top view of a sea urchin revealing the global plate arrangement **b** Close-up photograph of a sea urchin showing the differently sized plates

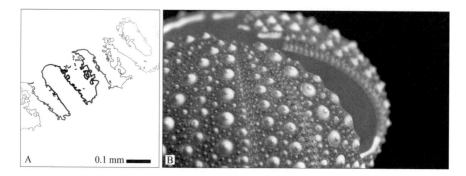

Fig. 3.3 Local load transfer **a** Schematic microscopic view of the plate's connection, revealing its finger-joint-like shape **b** Close-up photograph of a sea urchin with a broken test revealing the connection the neighbouring plates

Although global plate arrangements of echinoids have been analyzed, the developed performative architectural morphology does not transfer the patterns directly, as its main goal is the integration and interpretation of all manufacturing, architectural, structural and biomimetic principles. Instead, the subsequent pattern research needs to incorporate all the discussed principles, while also providing architectural qualities and the possibility to experience the structure in different situations. Therefore the developed design tool does not conclude in one fixed plate arrangement pattern, but rather opens up a variety of possible plate arrangements, whose diversity is mainly depending on a geometric research. Thus, all possible arrangement patterns have their own specific morphological space in which they can materialize and therefore be able to react on environmental conditions in their own field of responsiveness. The developed design tool is then able to automatically create the subsequent information chain needed for the manufacturing process, including the NC-code generation for all differentiated connection geometries.

4 Discussion

The considerable broadening of the range of possible plate structure articulations enabled by robotic manufactured finger-joined plates integrated into the developed computational design and manufacturing process as one design tool allows investigating different biological principles in relation to specific architectural characteristics. Following those introduced principles, the plate structure system can act as a continuous and rigid shell being able to divide, shape and even enclose spatial arrangements by connecting relatively small plates cut out of sheet material without any additional fastener. Using the principles of a double-layered system it is even possible to apply differentiated openings and include insulation while benefiting from the tethering's structural advantages (Figs. 4.1 and 4.2).

Ultimately the introduced methodology could be further investigated especially with regards to architectural design, being able to respond specifically and therefore in a differentiated, yet environmentally responsible way to a given set of system-external and system-internal conditions due to the developed design tools.

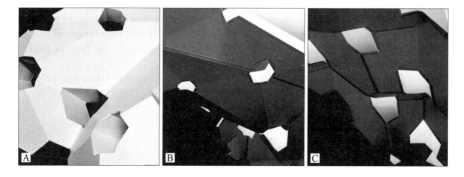

Fig. 4.1 Possible plate arrangements **a, b** Investigation of permeability and light modulation **c** A different plate arrangement during the same analysis

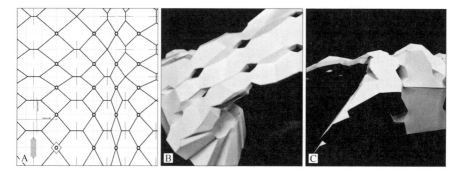

Fig. 4.2 Development of a performative plate arrangement **a** The plate structure is able to change the structural direction through its topology and perforation **b, c** The plate structure's possible articulation

Acknowledgements. This research project has been conducted in cooperation with the Institute of Building Structures and Structural Design, University of Stuttgart (Professor Jan Knippers, Markus Gabler, Alexander Hub and Frederic Waimer) as well has the Plant Biomechanics Group in Freiburg (Professor Thomas Speck and Dr. Olga Speck) as well as Christopher Robeller, Institute for Computational Design, University of Stuttgart and Markus Burger.

References

Barnes, R.D.: Invertebrate Zoology. Holt-Saunders International, Philadelphia (1982)

Buri, H., Weinand, Y.: Origami: Faltstrukturen aus Holzwerkstoffen. Bulletin Holzforschung Schweiz 2, 8–12 (2006)

Buri, H., Weinand, Y.: ORIGAMI - Folded Plate Structures. In: 10th World Conference on Timber Engineering, Miyazaki (2008)

Hensel, M., Menges, A.: Versatility and Vicissitude, An Introduction to Performance in Morpho-Ecological Design. In: Hensel, M., Menges, A. (eds.) Versatility and Vicissitude, Architectural Design, vol. 78(2), pp. 6–11. Wiley Academy, London (2008)

Nachtigall, W.: Bau-Bionik: Natur, Analogien, Technik. Springer, Berlin (2004)

Schindler, C.: Information-Tool-Technology: Contemporary digital fabrication as part of a continuous development of process technology as illustrated with the example of timber construction. In: ACADIA Conference 2007 Expanding Bodies. Proceedings of the International Conference, Halifax Nova Scotia (2007)

Schindler, C.: ZipShape – Gekrümmte Formstücke aus zwei ebenen Platten durch geometrisch variables Verzinken. Bulletin Holzforschung 15(1), 9–11 (2007)

A Technique for the Conditional Detailing of Grid-Shell Structures: Using Cellular Automatas as Decision Making Engines in Large Parametric Model Assemblies

Alexander Peña de Leon and Dennis Shelden

1 Introduction

The Steel GridShell of the Yas Island Marina Racetrack Hotel designed by Asymptote Architects was completed in November 2009, the integrated delivery of project team was comprised of Wagner Biro as the GridShell Fabricators, Schlaich Bergermann as the structural engineers and GridShell consultants, Front Inc as the Panel Assembly Fabricators, Evolute as the tessellation optimization consultants and Gehry Technologies Inc as the Integrated Delivery of Project consultants. In this paper we will focus on one aspect of the Integration Process: the automation process of generating a Digital-Mockup (DMU) of the Steel GridShell and Panel assembly, developed at Gehry Technologies.

1.1 Collaborative Framework

The Delivery process of the Yas island marina hotel grid shell was mediated through the use of a digital collaborative framework using Digital Project™, a Catia™V5 PLM solution for the architecture industry, the framework served as a centralized repository for the integration of fabrication specifications for manufacturing and assembly constraints. The collaborative framework enhanced the synchronization of concurrent engineering practices and the consolidation of heterogeneous data among dispersed teams and disciplines.

A detailed account of how the delivery process and collaborative framework was developed is outside the scope of this paper, this paper will focus primarily on the outcomes of an automation strategy for creating digital mockup assemblies of steel grid shells. The collaborative framework is mentioned to provide a context of where

Alexander Peña de Leon
Sial Royal Melbourne Institute of Technology, Melbourne, Australia

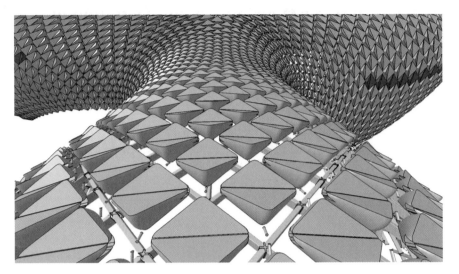

Fig. 1 Yas Island Digital Mock-Up, Highlighting panel clashes in red

the automation strategy of the grid shell DMU is positioned within the larger context of the integrated delivery process of the Yas Island Marina Race track Hotel.

1.2 Wireframes

The role of wire frame data is to allow an overview of how all components are distributed over the design surface. The wire frame can then assist in locating, orienting and sizing bottom-up isolated assemblies neglecting physical and geometrical constraints and the three-dimensional context surrounding the components across the entire project assembly, in other words the bottom-up context free assemblies are not aware of how they relate to the top-down assembly spatial layout.

Although wire-frame information is sufficient to communicate a limited set of design intentions it cannot be used for investigating the "assemblability" of the model, checking for clash-detection at different levels of representation using appropriate fabrication tolerances and clearances.

The model is an internal mechanism to provide sanity checks and visually debug information that would otherwise be difficult to manage.

The wire frame data was perfectly suited for the setup of linear element structural analysis of the grid shell elements, wire frame data was effortlessly translated into simple data schedules, providing Schlaich Bergermann with neutral representations of the same information embedded in the master model; this enabled a loss-less platform-agnostic process of transferring data across different tools.

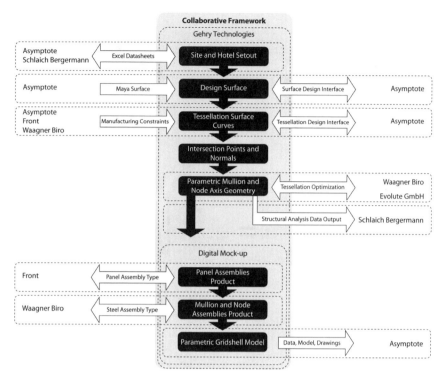

Fig. 2 Collaborative Framework supporting the Digital Mock-Up

1.3 Modelling Strategy

The spatial complexity of the Digital Mockup assembly, with in excess of 200,000 components within the assembly was a sufficient justification for developing a strategy for automating the massive instantiation of isolated bottom-up assemblies over the top-down wire frame assembly. Merging two conflicting modeling paradigms such as a top-down approach with a bottom-up approach can prove to be quite a challenging task within an associative geometric engine.

The Modelling Strategy and the Software Development strategy had to be coordinated effortlessly, providing a flexible framework for tackling the massive model assembly. Following the rationalization of the grid shell modeling strategy, the team was divided in two parts, the first developed hand modeled bottom-up assemblies representing each conditional detail of the grid shell as specified by the fabrication consultant, and the second part developed the computer code to instantiate each of the parametric assemblies the first team created. Both tasks had to be executed concurrently, even though both tasks are mutually dependent on one another. To solve the dependency problem, the two teams had to agree on the interfaces of the model and the interfaces of the computer code.

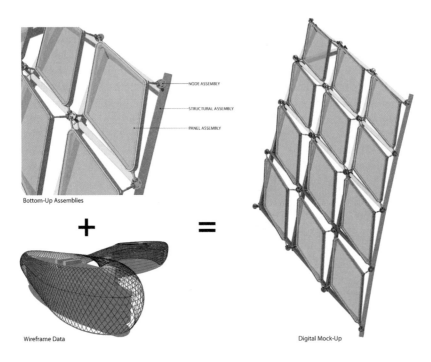

Fig. 3 Modelling Strategy for Digital Mock-up

These interfaces agreed upon before commencing any work did not change throughout the implementation phase of the modeling task or the implementation of the computer code for their propagation. Both the programming and the modeling of the components had loose implementations, as long as the inputs and the outputs of the components where kept the same, the framework would be intact.

2 Automation

The unsupervised instantiation of all components on the GridShell, begins with the execution of a visual basic script controlling Digital Project™ and Excel™, the code loops through each node and queries the existence map of the node neighbourhood, and with this map determining where the node in question is in relation to the grid shell, after identifying the relative location type (Top, Middle, and Bottom) the code queries the type map of the node neighbourhood to determine which rule has been met, if the condition matches any of the rules in the rule-set, the appropriate hand modelled bottom-up assembly is instantiated.

2.1 Search Strategy

The computational complexity of managing large assemblies is hindered by the speed of searching and manipulating records within large datasets. Understanding the computational complexity of algorithms and their relationship to the size of the datasets they manipulate, offers insights into developing a search strategy that renders the search time of the largest dataset equivalent to the search time of the smallest dataset.

The speed of searching an element within a massive dataset is proportional to the number of cycles required to find the element. An inefficient search strategy in its worst case scenario will take longer to search the dataset as the size of the dataset grows. The best case scenario is when the search time is always a constant time regardless of the size of the dataset.

Hash Maps are container structures that define the index of an element by mapping a key to a function that returns the object in linear time without needing to explore the dataset, and more relevant to this discussion is the facility of Hash Maps to query the existence of an element by asking whether the Key is contained within the Data set.

By maintaining the Vertices of the Grid Shell in Hash Maps, we can search any node in linear time regardless of the number of Nodes in the grid shell, searching for a node and its neighbors takes no longer than one cycle, computationally this means there is no speed penalty for querying about the existence of a node and retrieving its coordinates or any other pertinent properties.

If we follow the thinking process of hashing functions, the index of the vertex is mapped using an indexing pattern that encodes the topological spatial situation of the node in relationship to the grid shell. By reading the index of a node we should know how to locate the node within the grid shell and any of its surrounding members.

2.2 Node Mapping and Indexing Strategy

In the Yas Island Grid shell the virtual diagonal gridlines inclined forward where named *V lines* and the Back Inclined diagonal gridlines where named *H lines*. A Node in the grid shell is technically defined as the intersection point of a *H Gridline* and a *V Gridline*, so it follows that the vertex is named using the nomenclature ("H"+ H gridline number +"_" + "V" + V gridline number) for example a vertex indexed as *H200_V150* is located where Gridline *H200* intersects Gridline *V150*.

With a spatially aware indexing Nomenclature we can get the names of the neighbors of any node simply by decomposing the name of the node in question and adding or subtracting an offset value depending on the order of the gridlines. For example if we have node "H200_V150", we can extract all of its neighbors from the hash map simply by decomposing the name into two parts, first the H component 200 and second the V component 150, by adding or subtracting an offset value to the

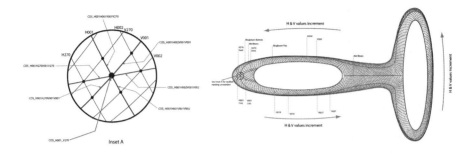

Fig. 4 Topological Naming Convention for Nodes, Beams and Panels

components we can get the name of the relative neighbor as many rows or columns away as the value of the offset we specified , since the grid shell is not infinite there is a possibility the new constructed name will not exist in the hash map of the nodes, by checking the existence of the 4 neighbors around a node, we can determine if a node is located at an upper edge, bottom edge or anywhere inside the grid shell not touching and edge.

The indexing and search strategy was universal to the project, different Hash Maps were used to cache the properties of each node, beam and panel. By combining the indexing and search strategy with maps of Structural Crossections, Node types and Panel Types, we could create a permutation of all typical conditions within the grid shell.

Prior to beginning the modeling task, the permutation of all explicit conditions were identified. Most of these conditions were already highlighted in the details drawing set provided by Wagner Biro of the different node types, beam types and support elements.

2.3 Cellular Automata as a Decision Making Engine

By identifying all of the explicit conditions within the project, we could anticipate all the place holder assemblies that needed to be modeled, and we could hard code into the propagation software a rule set for each condition managed by a cellular automata global function. Cellular automata are discreet models that simulate intelligence or decision making by using rule-sets to analyze a grid of cells, one at time, changing the current state of each cell as a function of the surrounding states in relation to its own state.

The cells can be abstracted to represent any concept that requires context aware decision making, in the case of the Yas Island Grid shell the cellular automata was a perfect mechanism for managing the unsupervised propagation of more than 200,000 components.

A Technique for the Conditional Detailing of Grid-Shell Structures

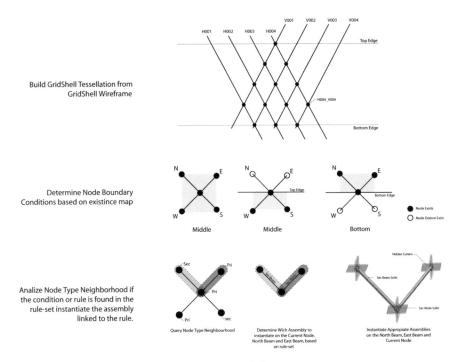

Fig. 5 Wireframe Analysis and Assembly Instantiation

3 Conclusions

When managing large model assemblies, high component counts are conveniently maintained by using wire frame data which is a Low level of Detail representation of a physical entity, these representations are not sufficient for investigating "assemblability" issues and misfits. A new model had to be built over the wireframe model, with a level of detail equivalent to the physical mock-up for conducting clash-detection, reachability, and swept volume analysis.

Our team devised a mechanism for the unsupervised creation of the digital mock-up instantiation of the Yas Island Marina Hotel using a novel application to the traditional cellular automata concept. The decision making engine was made possible by a high performance searching strategy, allowing fast lookups of data within the massive dataset controlling the nodes, beams and panels. The fast lookup was supported by an indexing heuristic that embeds geospatial and topological data in the naming of a component, allowing the retrieval of a nodes neighbourhood from basic string manipulations.

Parameterization and Welding of a Knotbox

Daniel Lordick

Abstract. A climbing frame called a Krabbelknoten (approximately: crawl-through knot) with a sophisticated mathematical background and constructed from approximately 600 meters of stainless steel wire, was created for the Erlebnisland Mathematik Dresden. We present the process of form finding, the parametric transformation of the surface into a framework of wires, the preparation of the welding in a traditional locksmith's workshop, the algorithmic layout of the supporting structure, the solution of a crucial point during the knot's transportation and last but not least, the integration of safety precautions into the design.

1 Location

The *Erlebnisland Mathematik* [4], established in 2008 at the *Technische Sammlungen Dresden* [8], is a permanent exhibition of mathematical objects. The aim of the exhibition is to get children interested in mathematics with the help of a playful environment and many hands-on exhibits. The continuous success of the exhibition led to an expansion of the exhibition space in February 2011. For the new space the curators wished to create a big object for physical interaction, especially climbing. The idea was to use the form of a knot to refer to the subject of knot theory. This idea finally led to the *Krabbelknoten* (Fig. 1 a).

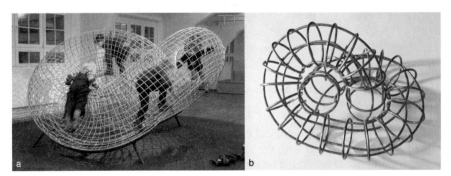

Fig. 1 a Children climbing through the Krabbelknoten in the Erlebnisland Mathematik Dresden, **b** First naïve model from the locksmith's workshop

Daniel Lordick
Institute of Geometry, University of Technology Dresden, Germany

2 The Shaping of the Knotbox

After some initial and naïve experiments by the locksmith's workshop [5] (Fig. 1 b) Prof. Ulrich Brehm at the TU Dresden came into play. He is working on topology and has developed sophisticated mathematical objects he calls *knotboxes* [2]. At first sight they seem to be simple but become increasingly attractive the more you get into this topic.

Starting point is knot theory. A mathematician's knot differs from everyday knots in that the ends are joined together so that it cannot be undone. It is a closed curve. Now according to Ulrich Brehm, a knotbox connected to a special knot can be described as a surface separating neighboring parts of the knot and forming one channel. Thus, the knotbox is a surface with two openings but without any edges. To better understand its topology we look at it from the result: if you imagine the simplest curve that can be drawn through the channel of the knotbox and connect the ends of this curve using the "outer" space of the surface, you would have the knot the knotbox had been derived from. This implies that one bight of the knot is always "outside" of the knotbox. But even if a knotbox looks like a compact object the term "outside" might be misleading, because a knotbox of a (single curve) knot is always a surface with only one side. It has the mathematical property of being *non-orientable*.

In preparation for the *Krabbelknoten* at the Erlebnisland Mathematik, let us take another approach to the idea of the knotbox: cut the curve of the knot open at one point and let the curve be a tube of some thickness. Now pump up the tube until adjacent parts of the tube melt into one surface and no other channels remain than those containing the original curve. Finally remove the two possibly overhanging parts of the surface resulting from the open ends of the original curve.

Fig. 2 a Thickened trefoil knot, **b** A stereographic projection of a minimal surface in S^3 [7], **c** A minimal rectangular knotbox

While a knotbox is a topological entity the actual form of a knotbox can be designed freely. Let us start with the simplest example of a nontrivial knot, the socalled trefoil knot (Fig. 2 a). The trefoil knot can be obtained by joining the two loose ends of a common overhand knot together, resulting in a knotted loop. Ulrich Brehm developed a variety of proposals of how to implement a knotbox for the trefoil, such as a minimal surface in four-dimensional space projected into three-dimensional space [6] (Fig. 2 b) or, easier to understand, a rectangular model of 2x3x4 cubic modules (Fig. 2 c). While the first proposal has very nice mathematical properties it is unsuitable for a climbing frame because of the great differences in the tube's diameter. The second one appears to be even less attractive because of the very loose correlation to the shape of a common knot.

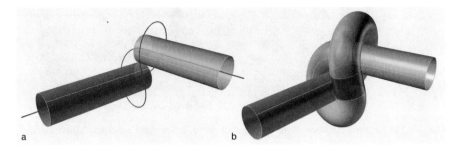

Fig. 3 a Cylindrical tubes touching under an optimized angle, smooth centerline of the knotbox with minimal length **b** Resulting pipe surface from five parts

That is why for the *Erlebnisland Mathematik* a different approach was chosen. We had to keep in mind that the surface had to be manufactured by a locksmith. So one meaningful restriction was to only use circles of equal diameter in one direction of the knotbox's channel. This essentially results in a pipe surface. So Ulrich Brehm started the design with a rounded string, tied in an overhand knot and tightened it as strong as possible. Thus the resulting path of the knot is as short as possible. Then he cut away the overhanging ends of the string. From this idea he developed a formula optimizing the centerline of the pipe surface. The result was a compound of two helical curves and three straight lines, where the straight lines are tangent to the helical curves (Fig. 3 a). The matching knotbox in essence is a compound of two helicoid pipes and three cylinders (Fig. 3 b).

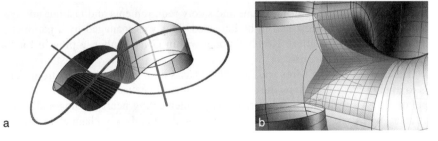

Fig. 4 a Blending surface between the cylinders, **b** Design of the NURBS surface (1st iteration)

But there is an extra part in the center of the knotbox, blending the cylindrical segments at the knotbox openings. This very special part is designed from a NURBS surface under the requirement to connect to the adjacent parts smoothly (Fig. 4 a, b). Finally the surface is smooth in all its parts, holds no edges but has

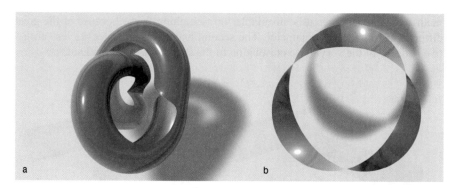

Fig. 5 a The knotbox cut open along the contact curve, **b** Standard 3-twisted Möbius strip

one curve of threefold contact that is tangential almost everywhere. If the knotbox were cut at this contact curve, it would not fall apart but represent a 3-twisted Möbius strip (Fig. 5 a, b). The edge of this strip again is a trefoil knot.

3 The Creation of the Framework

After the layout of the knotbox the resulting surface had to be transformed into a climbing frame of stainless steel. Thus it became necessary to apply a selection of parameter-curves, which are suitable for the realization with steel wires. The distance between neighboring parameter curves had to be chosen with a great enough width in order that no fingers could be trapped, this is more than 25 mm according to DIN EN 1176 [3], and smaller than 100 mm, so as not to form a trap for the heads of young children.

While the mathematical surface has zero thickness, the steel construction naturally does not. This resulted in countless points of conflict along the curve of contact, where manufacturing issues and safety requirements and last but not least also the aesthetic aspects had to be balanced. This was managed by a parametric representation of the entire Krabbelknoten within *Grasshopper*, a plug-in for the CAD software *Rhinoceros*.

The diameter of the wires was determined as 8 mm, which is rather strong. This is to avoid severe vibrations during use, which could cause fatigue to the welding points. Mostly there are two layers of wires intersecting nearly orthogonally. The circular wires are positioned at the inside of the knotbox's channel to provide good foot and handholds when moving through the pipe. But remembering that the surface has only one side, there had to be a line where the circular parameter curves had to switch from one side to the other. This was realized at the symmetry line of the knot, a short straight wire, where the adjacent wires end and no superposition exists (Fig. 6 a).

While there is only this short line where the thickness of the climbing frame is 8 mm, along the curve of contact the thickness of four layers of steel tend to total up to 32 mm. Moreover the wires at the curve of contact have different directions inducing a really complicated network of small triangles below the critical size mentioned above. To reduce these problematic points, it was necessary to synchronize the sequence of the circular wires with the 32 parameter curves parallel to the centerline of the knotbox. This had to take into account the dislocations induced by the offset of the wires from the original knotbox's surface. The last step was to systematically cut the wires before they actually reach the curve of tangential contact (Fig. 6 b).

Fig. 6 a Line of symmetry with a singular node, **b** Lattice in the area of the contact curve

4 The Welding of the Wires

Undoubtedly, the knotbox with all its smooth connections had to be fabricated identically to the computer generated design to guarantee the desired properties and aesthetics. But there was no automation possible, no digital design-to-production process. To achieve the required quality, several techniques were applied. At first the virtual model was given to the locksmith with the help of 3D PDF technology. So the craftsmen received a pretty good spatial understanding of the model and were able to clarify details of the superimposing wires.

The crucial point during construction was building the freeform part in the center of the knotbox. To help the locksmith to find the correct form and joins, we created a template from hardboard produced at the 3D LAB B25 [1] (Fig. 7 a). The template consists of interlocking section curves of the desired shape in two directions, which were cut by a laser cutter. Notches with a depth of 8 and 16 mm marked the positions of the wires. After the craftsmen had bent the wires according to the template they were able to weld them on top of the hardboard mask.

Fig. 7 a Welding template from hardboard, **b** All three cylinders connected to the central NURBS surface, **c** Adding circular and helical curves to the central body

Then the well-defined cylindrical parts were welded separately according to life-sized plans. Two of the cylinders fitted into the notches of the welding mask with one circle. This made precise connections possible. The third cylinder also has a well-defined position in relation to the three previous parts (Fig. 7 b).

To this body, finally the helicoid parts were added. At first some of the circular wires were attached to the body at the computed angle. Then three key helicoid curves followed. For each helicoid wire, the length had been computed and marked on the straight steel before bending. Then step by step, the helicoid curve was bent into the wire until the wire fitted according to the prepared connections of the torso. Now it was easy to fit in the missing wires by subdividing open parts of the network (Fig. 7 c).

So the Krabbelknoten is the result of in-detail engineering combined with very good craftsmanship.

5 The Algorithmic Layout of the Supporting Structure

The Krabbelknoten was developed for climbing through. So we had to define its position in space with the best balanced gradients of the centerline. It should not be too steep at any point. While turning the knotbox around on its line of symmetry we evaluated the rise of the centerline at various points. We stopped the movement when the gradient was as uniform as possible at a maximum of approximately 42 degrees. The next task was to fix the "cloud" of steel in exactly this position.

At first several prerequisites were conceived: the supporting structure should consist of tubular elements with only one diameter and hold the Krabbelknoten along helicoid curves. The stands should be orthogonal to the knotbox's surface and with their lower end sit on a triangle of rectangular beams (Fig. 8 a). This led to a setup for a grasshopper canvas where several parameters could be varied to achieve a supporting structure that was strong enough but as lean and as elegant as possible. With grasshopper the necessary plans were also generated, as well as the data for the laser cutter to produce some life size templates (Fig. 8 b, c).

When the locksmith saw the plans he stated that he could not bend some of the circular curves. His tool for tubes is limited to only one special radius. Luckily it was no problem to adapt to this restriction, because grasshopper generated the new plans and masks automatically.

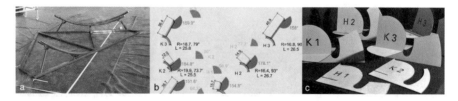

Fig. 8 a Supporting structure of the Krabbelknoten **b** Real-time updated plans, **c** Welding templates from the 3D LAB B25 [2]

6 The Solution of a Crucial Point during Transportation

The Krabbelknoten had to be transported from the workshop to its destination in one piece. Because the exhibition space is on the third floor of an old factory building, some problems occurred. The knot had to be lifted by a mobile crane and pulled in through a window, which a carpenter had removed especially for this purpose. But the Krabbelknoten had still not yet reached its destination but had to pass through another doorway. What we knew was, the minimum bounding box of the Krabbelknoten would not fit through this door, no matter how the Krabbelknoten is turned. Nevertheless, virtual simulations had offered hope that the knot might pass through the doorway. To finally clarify this, models of the doorway and the knot were built to a scale of 1:20 (Fig. 9). And indeed, through a special twist of the Krabbelknoten while moving through the doorway, it fitted. The transport has been documented in a video.

Fig. 9 Sequence of the testing weather the knotbox does or does not fit through a special doorway in the Erlebnisland Mathematik [1]

7 Final Remarks

During the whole designing and building process, strict safety regulations for playground equipment had to be considered. Although many conflicts had been eliminated with the help of the parametric design of the wires, one essential problem had to be corrected manually. Along the curve of contact the wires inevitably form pinch points. These pinch points are dangerous because they could trap cords on children's clothing. To amend this, synthetic compound material has been filled into the interstices. This measure interferes with the purity of the steel construction but transforms it from a 200-kilogram model of a knotbox into a usable toy that since the reopening of the Erlebnisland Mathematik in February 2011 is one of the main attractions for thousands of children.

References

[1] 3D LAB B25, http://www.math.tu-dresden.de/3D-LAB-B25 (accessed June 10, 2011)
[2] Brehm, U.: Polyhedral surfaces (extract of a talk). In: Proceedings 44/1989, Mathematisches Forschungsinstitut Oberwolfach, Geometry, October 15-21, p. 2 (1989)
[3] DIN Deutsches Institut für Normung e.V. Taschenbuch 105, 6th edn., Beuth Berlin, Wien, Zürich (2009)
[4] Erlebnisland Mathematik Dresden,
http://www.math.tu-dresden.de/alg/erlebnisland
(accessed June 10, 2011)
[5] Hesse, R.: The Locksmith's Workshop, http://www.edelstahl-und-design.de (accessed June 10, 2011)
[6] Lawson Jr., H.B.: Complete Minimal Surfaces in S3. The Annals of Mathematics, Second Series 92(3), 335–374 (1970),
http://www.jstor.org/stable/1970625
[7] Mathematical Models Collection, Institute of Geometry, University of Technology Dresden, http://www.math.tu-dresden.de/modellsammlung (accessed June 10, 2011)
[8] Technische Sammlungen Dresden, http://www.tsd.de (accessed June 10, 2011)

Viscous Affiliation - A Concrete Structure

Martin Oberascher, Alexander Matl, and Christoph Brandstätter

Abstract. This paper describes the design criteria and the construction methods used to construct a free formed concrete roof structure. The innovative construction method refers to the competencies and teaching focus of the Academy in the field of concrete application. The generated geometry is based on a set of fluid simulations.

Fig. 1 Rendered view from outside, **Fig. 2** Interior view from Physical model

Martin Oberascher · Alexander Matl
Soma-Salzburg

Christoph Brandstätter
Brandstätter ZT GmbH

1 General Information

The Building Academy Salzburg is a training centre of the local building sector. The Chamber of Commerce, section building trade as the owner and the Building Academy as the operator of the complex, are keen to establish, an attractive cultural event space alongside the technical training units. In the past the building complex has hosted classical concerts and performance, as well as sprayer, breakdance and hip-hop Battles for the young generation. Beyond that, during the summer school holidays, the spaces are used as rehearsal rooms for the Salzburg Festival. The given task was to implement an extended foyer as a representative meeting zone for the existing building complex. Moreover the aim was that the new foyer should connect the rebuilt multi-purpose hall with the school. soma's design proposal was to create a unique architectural intervention in the existing building structure that supports the different stakeholders utilisation.

2 Architectural Intention

Cross-links between the application areas of cultural concerns and technical schooling are not always obvious but bringing these two stakeholders closer together is the intention of the architectural concept. While the time frames of the usage by the students and the audience are not congruent, nevertheless the same demands on the foyer exist. In order to give this blending a visual representation, atmospheric uniqueness and structured surfaces became the important issue. The entrance structure establishes a consistent and smooth transition between the exterior building volumes and the interior operation sequences. The new roof structure extends into the existing building and creating an intuitively understandable guidance system and a new functional zoning. The transition between the pending box (seen from outside) and the interior roof structure should be read as affiliation between the existing building typology and contemporary spatial concepts.

3 Form Generation

To generate the structure a simulation of fluids based on particle simulations was used. In this case the simulation was set up with a software called realflow. Liquids have three essential parameters: viscosity, density and surface tension. The interactions between these three physical properties have been tested on the computer in a series of variations. The goal was to generate a pattern with a high proportion of holes. In the simulation this viscous liquid is observed in a three dimensional digital model of the existing building to get a material flow mapping of a self organizing structure. In order to arrive at a guidance system, a functional zoning, and also to meet structural requirements a series of force emitters and attractor fields is applied to the scene. These vectors affect the fluid in such a way that it contains all the required architectural information. After the form generation was finished the geometry have been converted to the 3D nurbs modeling software Rhino 4.0 which has been used for all the technical approaches.

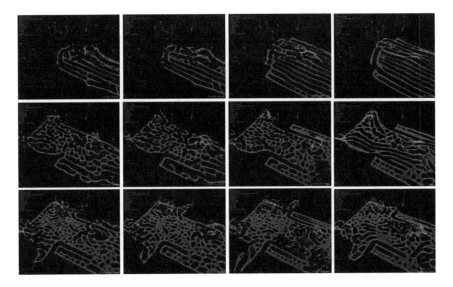

Fig. 3 Simulation frames with different liquid parameters

4 Surface Exploration

In order to optimize the production process and to insure the quality of the final materialization a series of structural surface models, including a 1:1 real-scale mock up has been produced and tested. The focus of investigation has been on the physical production of the moulds. The density and structure of the CNC milling paths have a significant impact on the lighting effects of the finished concrete surface. Furthermore the tessellation method and mesh size have also been a particular focus of our examination.

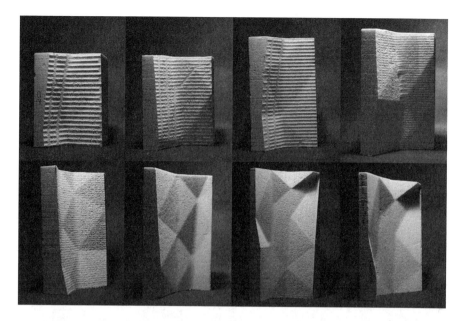

Fig. 4 Mould examination for surface finishes

5 Innovative Construction

The complexity of the design with its static requirement is far beyond a conventional reinforced concrete structure. The technical challenge is the translation of the free-form body in a cast in place concrete structure fulfilling the static requirements of the roof. Due to the complexity of the surface a structural solution cannot be found using a conventional reinforced concrete structure. A composite system of steel hollow profiles and cast in place concrete turned out to be most efficient and practicable. The high structural strength of steel combined with concrete's compressive strength and its ability to form complex surfaces are ideal for the task. Composite structures have recently acquired increasing importance mainly in multi-storey buildings and bridge constructions. Steel hollow profiles are cast in the free form concrete structure. To achieve sufficient bond of the steel members with the concrete, steel studs are welded onto the profiles. As they follow the surface, all steel studs must have a different length. The structure consists of a beam grillage (roof) which is supported by the existing building and by two branched composite columns. To ensure a minimal crack width under load a steel reinforcement net is provided in the vicinity of the concrete surface.

Furthermore the structure has to be tested in 1:1 mock ups, applying the maximum load to observe crack behaviour and crack visibility of the finished concrete surface.

Viscous Affiliation - A Concrete Structure

Fig. 5 (Mock Up) CNC-method, **Fig. 6.** (Mock Up) styrofoam mould'

Fig. 7 (Mock Up) embedded steel grillage, **Fig. 8.** (Mock Up) hollow profiles with steel studs

Fig. 9 (Mock Up) Finished real scale model, **Fig. 10.** (Mock Up) Load test

6 Assembling

The free form structure is directly translated into CNC milled styrofoam formwork. The steel girders with the applied net reinforcement is embedded in this formwork. To ensure a constant concrete cover of 3 cm the reinforcement bars are assembled in a second styropor form which is 3 cm narrower then the final body. This net of reinforcement bars consists of stirrups and longitudinal bars. For precise assembling on the site, the geometry of the reinforcement and the steel studs are defined by a 3-dimensional computer model which is provided to the contractors.

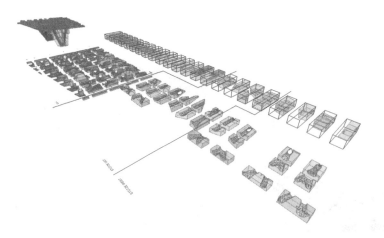

Fig. 11 3d model mould assembly

The following criteria were significant for the art, technical, and financial supervision:

- precise and current 3d model which contains all the information of all the different manufacturing companies (including reverse engineering by a geometer).
- Logistical coordination and processing.
- Material properties of the raw cnc-milling material (maximum dimensions, density, minimum waste).
- Tool properties of the cnc-milling machine (maximum traverse path, milling-bit, milling-depth, step-sizes, etc.
- Mesh density in correlation to the milling paths.
- Formwork finishes in the focus of exposed concrete and visibility of cracks.
- Positioning of crack reinforcement on a freeform shape.
- Cracking behaviour and crack pattern (existing technical standards where hardly applicable in a field of experimenting cnc-milled formwork).
- workability properties of self compacting concrete (SCC).
- Assembling of steel structure, head studs, crack reinforcement and formwork.

Fig. 12, 13 (realization) reinforced Columns with styrofoam formwork, **Fig. 14** (realization) Steel grillage on the crane

Fig. 15 (Realization) grillage with steel studs and mounted crack reinforcement, **Fig. 16** (realization) Embedded steel structure in the assembled styrofoam moulds

Fig. 17 (Realization) Pouring the concrete

7 Conclusion

Realizing this structure in exposed concrete posed major challenges in the area of form work construction which could only be optimized in 1:1 mock ups. The specially developed manufacturing technique and the interdisciplinary production are essential to carry out such a project successfully. The high number of tests and experiments take account for the fulfilment of architectonic parameters to create a unique atmospheric impression. The result of the finished concrete structure will be seen after stripping the form on 20th of June.

Dynamic Double Curvature Mould System

Christian Raun, Mathias K. Kristensen, and Poul Henning Kirkegaard

Abstract. The present paper describes a concept for a reconfigurable mould surface which is designed to fit the needs of contemporary architecture. The core of the concept presented is a dynamic surface manipulated into a given shape using a digital signal created directly from the CAD drawing of the design. This happen fast, automatic and without production of waste, and the manipulated surface is fair and robust, eliminating the need for additional, manual treatment. Limitations to the possibilities of the flexible form are limited curvature and limited level of detail, making it especially suited for larger, double curved surfaces like facades or walls, where the curvature of each element is relatively small in comparison to the overall shape.

1 Introduction

Complex freeform architecture is one of the most striking trends in contemporary architecture. Architecture differs from traditional target industries of CAD/CAM technology in many ways including aesthetics, statics, structural aspects, scale and manufacturing technologies. Designing a piece of freeform architecture in a CAD program is fairly easy, but the translation to a real piece of architecture can be difficult and expensive and as traditional production methods for free-form architecture prove costly, architects and engineers are forced to simplify designs. Today, methods for manufacturing freeform concrete formwork are available, and more are being developed [1, 2, 3, 4]. The common way of producing moulds for unique elements today is to manufacture one mould for each unique element using CNC milling in

---

Christian Raun · Mathias K. Kristensen
ADAPA-Adapting Architecture, Denmark

Poul Henning Kirkegaard
Department of Civil Engineering, Aalborg University, Denmark

cheaper materials, but since the method is still labor intensive and produces a lot of waste, research is carried out in several projects to find a solution, where one mould simply rearranges itself into a variety of familiar shapes. Such a concept has natural limitations, but would become a complimentary technology to the existing.

The present paper describes the development of a digitally controlled mould that forms a double curved and fair surface directly from the digital CAD model. The primary motivation for the development of the mould is to reduce the cost of constructing double curved, cast elements for architecture, both in-situ cast and modular. Today, such elements are usually cast in milled formwork that is expensive and produces a lot of waste. Architects are often limited in their freedom of design by the high costs of the existing methods and as a result, the possibilities for drawing and evaluating complex shapes in architecture today, are not reflected in the build architecture.

2 Concrete Casting Techniques

Today, a number of technologies have emerged, that offers casting methods for a range of purposes. On a large scale, the market is dominated by well known techniques such as precast elements made from standard moulds and in-situ casting in standardized modular systems. On a small scale, new methods for casting and new types of moulds have emerged to meet the rising demand for customization and creation of curved concrete architecture. Some of the methods for double curved moulds which have been investigated related to the present project are mentioned below.

2.1 Milled Foam Moulds

The milled foam method represents the newest and the most economic version of custom manufactured moulds, historically made by hand and recently milled in different materials using CNC.

The advantage of foams in comparison to heavier materials is, that they are cheaper compared on volume, they allow fast milling, and they are easy to manually alter and fair after the milling process, that leaves a grooved surface texture. The main strength of the method is that it can be used for very advanced geometry as long as it is possible to de-mould the casted object. Further there is almost no curvature or detailing level limitations besides that of the milling tool. Another clear advantage for this method is that the entire surface is manufactured to tolerances. The weakness of the method is, that it requires manual fairing and coating to a large extend, if the surface has to be of a perfectly smooth, polished quality. For a large project, the formwork is extensive, and after use it has to be thrown out, creating even more waste than was produced during milling.

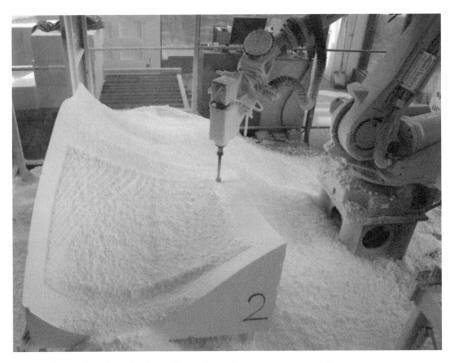

Fig. 1 Photo of a robot CNC-machine milling in a styropor material

2.2 System Based Traditional Formwork, PERI

PERI is a German producer of traditional scaffolding systems, but they have expanded their product portfolio to include both flexible single curvature formwork and custom double curved formwork. PERI specialty is that they use standard components for the production of all their form work and both the single curvature flexible form and their custom double curved forms are integrated into a complete and rationalized in-situ system. They have also developed software that can automatically determine what parts are needed based on a given geometry. It is a complete and reliable solution from software to hardware, design to construction. The main weakness is that there is still waste produced in the process of creating the double curved moulds, and that it is only possible to create double curved surfaces of very small curvature.

The systems and methods shown above cover each their different aspects of freeform architecture. Whether building scale or curvature is taken into consideration, there seems to be a gap in scale from the milled moulds to PERIs large scale buildings. PERIs boards or plywood sheets forced to create double curvature has a fairly small maximum curvature, and it seems futile to use a precision tool like a CNC milling machine, with its capability to produce very accurate and complex geometries, to create larger modules of relatively small curvature without further

Fig. 2 Photo showing PERIs single curvature scaffolding

detailing. When looking at this curvature scale - smaller buildings created from a larger number of precast elements of familiar scale and curvature, it seems such elements could be generated from a common tool, the curvatures of which could be found between the maximum curvature of the force-bend scaffolding from PERI and the small, complex curvatures possible by milled moulds. A flexible tool could be competitive with foam milling in this area, if it were made, so that no additional, manual treatment of cast elements or surface were needed, no waste produced and production speed in comparison to equipment price were better. A tool for creating modular solutions should come with a software or system to rationalize production and communicate possibilities to architects. At the same time, the direct connection between drawing and machine, as with the CNC miller, should be established, to get an automated process. If a tool can be created to meet the criteria stated above, it could help promote the construction of the freeform architecture that is so commonly seen in digital architecture and competition drawings today, by offering cheaper and more efficient custom building parts. It could help bring the build architecture closer to the digital possibilities.

2.3 Flexible Moulds

The most important aspect to consider when designing a flexible form is its limitations. The wider the desired range of possible shapes, the more difficult and advanced the construction will be. As discussed in the previous section, CNC milled foam moulds will at some point of complexity be the most attractive solution, as they are able to mill shapes that would be extremely hard to achieve by any other way of manipulating a surface. It is also clear, that no matter how a surface is manipulated in a flexible form, the very nature of the method results in a specialization in a common family of shapes. For instance, if a flexible mould were to create a perfect box, it may be designed to take different length, width and height, but because it needs specialized geometry like corners, it would be unable to create a sphere with no corners. A flexible mould aiming at the ability to do both, would possibly fail to achieve a perfect result in either case.

It all comes down to the fact, that every point on the surface of a flexible mould does not have the ability to change from continuity to discontinuity, because that would demand an infinitely high number of control points. Without an infinitely high number of control points, the flexible form, therefore, has to aim at creating smooth, continuous surfaces, the complexity of which must simply be governed by the number of control points. It is then left to decide, what the least number of control points is, relative to the properties of the membrane which will result in a mould design capable of achieving the curvatures needed for most freeform building surfaces. The initial motivation for the design of a flexible mould for double curved surfaces, was the encounter of other attempts to come up with a functional design for such a form, and the market potentials described in these projects. The technical difficulties and solutions defined in the projects presented here have been the inspiration for our present design.

2.4 Membrane Mould

The mould concept is to have a flexible membrane manipulated by air filled balloons. The use of balloons solves the problem of creating smooth bulges on a membrane with no stiffness in bending, but it is hard to control the tolerances. The edge conditions are, however, defined relatively precise by linear, stiff interpolators connected to rods and angle control.

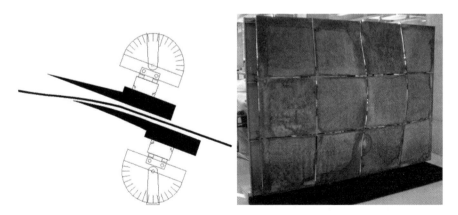

Fig. 3 Edge control by angle measurements [1]

Fig. 4 This edge control means that the panels can be joined to create a relatively continuous surface [1]

2.5 North Sails

North Sails in North America produces custom cast sails in a digitally controlled, flexible form that uses a principle, where stiff elements created a smooth surface between points defined by digitally controlled actuators. They simply use what appears to be a thick rubber or silicone membrane which has an even surface, since it is supported by a large number of small, stiff rods placed close together underneath it. The small rods are placed on top of larger rods connected to the actuators. This simple system is possible because of the relatively small curvature in comparison to the mould size. The mould is highly specialized and appears to have been extremely expensive, but it is the best example of a flexible mould concept, that could easily be used to cast concrete panels, and it has been the main inspiration for the principles used in our mould.

Fig. 5 North Sails mould with numerous actuators

3 Concept for a Dynamic Surface System

In the proposed dynamic mould system, where only a set of points is defined, a stiff membrane interpolates the surface between those points. Stresses in the deflected membrane will seek to be evenly distributed and therefore it will create a fair curve through the defined points. A stiffer member will have a more equally distributed curvature, while a softer member will tend to have higher peaks of curvature near the defined points. This relation between the physical properties of a stiff member and the mathematical properties of a NURBS curve can be applied to surfaces as well. If a plate interpolator can be made, that has an equal stiffness for bending in all directions, and the freedom to expand in its own plane it would constitute a 3D interpolator parallel to the well known 2D solution. To function as a surface suitable for casting concrete or other substances against without the need for further manual treatment, the membrane should be durable and maintain a perfectly smooth and non-porous surface as well. A membrane with these properties has been developed for this project, and it is the core of the dynamic surface mould invention. The number of actuators in a row defines the precision and possible complexity of the surface. A smaller number of actuators require a stiffer membrane and less control, a larger number means softer membrane and better control. In this way, the amount of actuators needed depends on the complexity of the surface.

Fig. 6 Illustration of a surface deformed by 3, 4 and 5 actuators

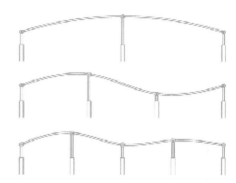

Five actuators in a section have been chosen not only because of the finer control, but also because the coherence between the NURBS surfaces in a CAD drawing and the physical shape of the membrane is better. Smaller leaps between the pistons mean less deflection caused by the viscous pressure, and most important, the edge conditions in a 5x5 configuration is less affected by the deflection elsewhere on the membrane, than they are with the 4x4.

3.1 Functionality and Limitations

The mould can take any digitally defined shape within its limitations within one minute from the execution of a program reading 25 surfaces coordinates directly from the CAD design file. Once the actuator pistons have taken their positions, the handles must be adjusted manually to a fixed angle calculated by the program. The main limitation of the mould is its maximum curvature. It is defined by the construction of the membrane, and for the prototype, it is approximately a radius of 1.5m. The system can be scaled to achieve smaller radii. Another limitation is that the surface designed has to fit within the 1.2m x 1.2m x 0.3m which is the box defined by the pistons. This box is adequate to create a square piece of a sphere as big as the mould, with a radius of 1.5m. For most of the freeform architectural references, these limitations mean, that it is conceivable to produce the main parts of the facades. Control of the actuators via CAD software is programmed using the Arduino platform and Rhino supporting NURBS which is ideal for generating surfaces applicable to the mould system. The information layout is based on the following; Arduino informs Rhino about current positions of all actuators and Rhino issues commands to Arduino which then positions all actuators. Controlling a surface in Rhino generates real-time feedback in the mould system and in-program information about curvature degrees and possible warnings. Typical scenarios of use have been implemented as to help from an early point in a design process. Starting from at double curved wall where the program will issue information about different subdivision possibilities and possible problems relating to manufacturability, to the coordinated virtual and physical handling of the separate elements on and off the mould system.

Fig. 7 Casting of a double curved fiber reinforced concrete panel

The following illustrations explain how the flexible mould can be used from design to production of freeform architecture.

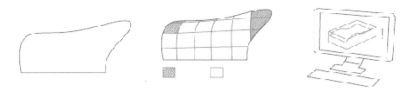

Fig. 8 Double curved surface, a subdivision of surface and validation of subdivision

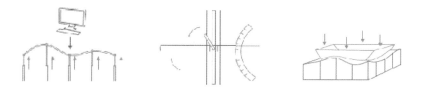

Fig. 9 Positioning of actuators, adjusting edge angles and mounting mould sides

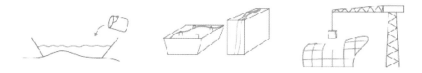

Fig. 10 Pouring filling, single or double sided moulds and of mounting elements

Fig. 11 Surface made up of elements produced with the double sided mould technique

4 Conclusion

Complex freeform architecture is one of the most striking trends in contemporary architecture. Today, design and fabrication of such structures are based on digital technologies which have been developed for other industries (automotive, naval, aerospace industry). The present paper has presented traditional production methods available for free-form architecture which force architects and engineers to simplify their designs. Further the paper has described the development of a flexible mould for production of precast thin-shell fiber-reinforced concrete elements which can have a given form. The mould consists of pistons fixing points on a membrane which creates the interpolated surface and is fixed tothe form sides in a way that allows it to move up and down. The main focus for the development has been on concrete facade elements, but a flexible, digitally controlled mould can be used in other areas as well. Throughout the project interest has been shown to use the mould for composites as well, and among other ideas are the idea of casting acoustic panels, double curved vacuum formed veneer and even flexible golf courses. The flexible mould concept has the ability to advance modern, free form architecture, as the concept offers cheaper and faster production of custom element.

Acknowledgements. The present research is support by the *Det Obelske Familiefond*. The authors highly appreciate the financial support.

References

1. Pronk, A., Rooy, I.V., Schinkel, P.: Double-curved surfaces using a membrane mould. In: Domingo, A., Lázaro, C. (eds.) IASS Symposium 2009, Evolution and trends in design, analysis and construction of shell and spatial structures, pp. 618–628 (2009)
2. Helvoirt, J.: Een 3D blob huid Afstudeerverslag 3370, Technische Universiteit Eindhoven (2003)
3. Boers, S.: Optimal forming, http://www.optimalforming.com (checked on the of June 30, 2010)
4. Guldentops, L., Mollaert, M., Adriaenssens, S., Laet, L., Temmerman, N.D.: Textile formwork for concrete shells. In: IASS Symposium 2009, Evolution and trends in design, analysis and construction of shell and spatial structures, pp. 1743–1754 (2009)

More Is Arbitrary
Music Pavilion for the Salzburg Biennale

Kristina Schinegger and Stefan Rutzinger

Abstract. This paper describes the design criteria for a temporary art pavilion and the development of an irregular complex structure with emergent load bearing capacities. Emergence is understood as the production of qualities out of quantity. This implies that - at least in the architectural context - the discussion about performance cannot omit the experiential.

Fig. 1 Temporary music pavilion / F.Hafele

Kristina Schinegger · Stefan Rutzinger
Soma, Vienna

1 Architectural Task

1.1 General Information

The temporary pavilion will create a unique presence for contemporary art productions in Salzburg, a city known predominantly for classical music. The main user of the pavilion is the Salzburg Biennale, a contemporary music festival. soma's proposal was chosen as the first prize winner in an open, two-stage competition in October 2010. The pavilion was erected for the first time in March 2011 for a period of 3 months and housed various events such as contemporary music concerts, video screenings and exhibitions. During the next decade it will be used for various art events at different locations. The structure can be divided into individual segments. By combining these segments in different ways or by reducing their number, the pavilion can adapt to its location. The removable interior membrane and the adjustable floor increase the flexibility of use.

1.2 Architectural Concept

Art is a cultural process involving many participants within a discourse. This process does not reveal itself at first sight, but unfolds through encounter and engagement. The pavilion's appearance emphasises this idea. It provokes curiosity and invites visitors to encounter the unknown and unusual.

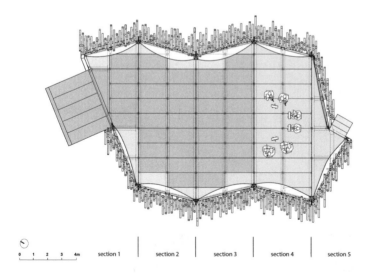

Fig. 2 Ground plan / soma

Fig. 3 Production / Unterfurtner

The architectural concept is based on a theme that is inherent to architecture as well as music – rule and variation. The design process of the pavilion is based on a simple repetitive element, a set of rules for aggregation, and the definition of the architectural effects aimed at. The single aluminium profiles with a uniform length produce an irregular mass-like conglomerate that changes its appearance during the day, according to the different light conditions. The structure allows an ambivalent reading as single members and as a merging whole, depending on the distance it is viewed from. The speculative intention behind this "obliteration" of the pavilion's structure is to prevent any conventional notion or cliché of "construction". Instead the pavilion should invite visitors to come up with their own associations and interpretations.

2 Structure

2.1 Framing Conditions

Thanks to computation complex structures employing disorder and randomness can be created and controlled. Although these irregular patterns are often applied to special building parts like facades applications for load bearing structures are still an exception. Furthermore irregular complex structures are often based on highly individual components.[1] The bottom-up strategy of the music pavilion is based on a repetitive linear base element that does not change shape. Likewise the reference surface (inner membrane) is a rather simple geometry in order to display

the complexity of the aggregation rules as effectively as possible. Furthermore the aluminium profile is cut from stock ware (6 m length) to avoid leftover material.
In the conceptual stage of the design process a set of frame conditions were defined and tested:

- the pavilion should appear as a mass not as a form
- the structure should consist of one simple repetitive element
- the distribution of members should be irregular while still showing a homogeneous density
- adjacent layers of the sticks follow opposite directions

The overall structural system of the pavilion is divided into 5 individual sections to increase flexibility of use. Each section consists of 20 vertical construction layers with a spacing of 20cm (start and end section have fewer layers). On each layer intersection curves with the reference surface will host starting points for the structural members. The distribution of points and positioning of the structural members takes place within a range of randomized distances and angles but at the same time prevents intersections. Due to individual positioning of members along each section curve, projection intersections with adjacent layers are generated. This process produces an interconnected structure.

The structural optimization by Bollinger Grohmann Schneider ZT GmbH takes the design rules above into account but also considers working loads, amount of connection elements and the maximum deflection of each segment. To evolve a

Fig. 4 Numbering / BGS ZT GmbH

structure *Karamba* (developed by Clemens Preisinger in cooperation with BGS Engineers and Structural Design Institute, Univ. of Applied Arts Vienna) was applied within *Grasshopper*. Combined with a genetic algorithm the optimized solution was filtered out of the multiplicity of solutions through combination, selection and mutation over many generations.

"The elements are aligned iteratively and interact in a parallel way. By repetition of the same calculation step and with the feedback of the results, the system is incrementally evaluated until a certain target value is reached or the system converges to a threshold value. Multiplicity denotes the simultaneous and parallel observation and adjustment of the individual elements in a single step. The coactions of multiplicity and iteration result in the system's ability to adapt to a given task." (Moriz Heimrath and Arne Hoffmann, BGS ZT GmbH).

2.2 Optimization as a Balancing Process

Optimisation is here understood as enhancing structural performance within architectural parameters and aesthetical intents. In addition to formal and structural aspects the amount of members is minimized without loosing the mass-like appearance.

The parametric model based on *Grasshopper* and *Karamba* enabled the architects and the engineers to simultaneously design and evaluate the structure. This process cannot be considered as a strictly parametric straightforward design generation process, but is rather a negotiation between architectural aspiration, structural behaviour, buildability, logistics of assembly, and cost control.

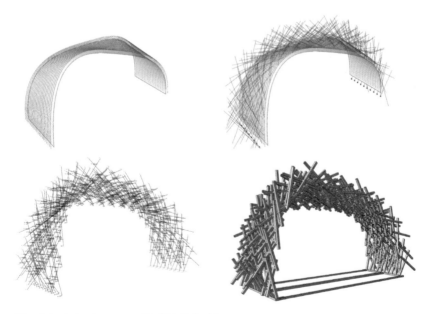

Fig. 5 Optimized structure / BGS ZT GmbH

Therefore structural performance cannot be considered as the initial generator for the structure, although it was crucial to embed the potential for structural performance already in the conceptual stage of the project.

3 Theoretical Context

3.1 Irregularity as an Aesthetic Principle

When designing irregular or random pattern the set up of framing conditions is decisive for the outcome. In case of the music pavilion the random factor was applied to prevent repetition and to produce a certain effect of mass and disorder. The guiding geometries and the set of rules provide the frame for randomness to appear.

Sean Hanna refers to the use of randomness within a targeted process as stochastic "in the real sense in that each of these rules is an aim or target and a level of unpredictability is able to enter the picture as long as the result comes sufficiently close to meeting it."[2] He sees a "new aesthetic" on the way in which order is not something considered as pre-given but emergent, driven by an interest in underlying principles in nature.

This tendency towards the design of rules and display of inherent principles is also a shift from an interest in *external form* towards the *inner logic* [3] or, as Stan Allen puts it, from *objects* to *fields*. In *fields* figures and forms appear through the observing eye of the beholder rather than being conceived and represented by the author. The designer rather creates a potential for the figure as a local "effect" within a heterogeneous field:

"What is intended here is a close attention to the production of difference at the local scale, even while maintaining a relative indifference to the form of the whole." [4]

Allen calls these *fields* "systems of organization capable of producing vortexes, peaks, and protuberances out of individual elements that are themselves regular or repetitive." He highlights the "suggestive formal possibilities" and the questioning of conventional top-down form controls. In his opinion fields also have the potential to provoke a re-addressing of use: "More than a formal configuration, the field condition implies an architecture that admits change, accident, and improvisation." [5]

3.2 Performative Structures

Irregularity can be understood as an aesthetical principle that combined with evolutionary computation can evolve emergent and *performative* qualities, as the vast field of ongoing architectural experimentation shows. Achim Menges and Michael Hensel have lately called upon architecture to rethink the hierarchical succession of the form-finding and materialising processes and proposed an integral way of evolving emergent structures by inherent logics of material systems and environmental influences. They suggest that analysis or simulation tools become generative drivers for the design process of *performative* structures.[6]

If emergence is understood as the production of certain qualities out of quantity [7] then - at least in the architectural context - this includes by definition the experiential or what has lately been discussed within contemporary practice as "subjective effects".[8] Following Stan Allen we should ask what this *more* can mean for architecture, beside spatial experimentation and the innovative rethinking of conventional building and materializing techniques.

Fig. 6 Perspective / F.Hafele

Fig. 7 Perspective / F.Hafele

Somol and Whiting's definition of *performative* in their appraisal of a post-critical practice opens a wider scope and a broader discussion. In contrast to the *difficult* and elite attitude of critical architecture that calls upon visitors to unravel complicated concepts, the *projective* practice stands for the *easy* and inclusive that puts an emphasize on atmospheric effects and spatial experiences. Rather than limiting performance to measurable objective operations, it includes speculation about experiential and even social effects of architecture.

The success of the term might be rooted exactly in this broadness and vagueness nature of its definition, yet the discussion could widen the scope of *material practice* from a quasi-scientific bottom-up experimentation towards contextual architectural implementation.

4 Conclusion

The design process of the music pavilion is actuated by the set-up of rules and framing conditions that could be understood as the inherent logic of the emerging structure. Nevertheless the experiential qualities and the external expression of the structure remain a principal focus of the architectural design. The mass-like appearance aims at underlining the creative character of our perception, since our brains are constantly trying to distinguish figures and patterns within disorder. Bottom-up means here also a speculative intention: Rather than to represent forms or meanings, the architecture produces an ambiguous mass to allow visitors come up with their own interpretations and associations. In this way the pavilion could be called *performative*. It wants to trigger engagement with contemporary music, not by being complicated or difficult but by displaying complexity in a playful way.

References

[1] Scheurer, F.: Turning the Design Process Downside-Up. Self-organization in Real-world Architecture. In: Martens, B., Brown, A. (eds.) Computer Aided Architectural Design Futures, Springer, Vienna (2005)
[2] Hanna, S.: Random has Changed. Archithese, 40–44 (June 2010)
[3] Frazer, J.: An evolutionary Architecture. AA Publications, London (1995)
[4] Allen, S.: From Objet to Field. In: AD, vol. 67(5/6), pp. 24–31 (1997)
[5] Menges, A., Hensel, M.: Versatility and Vicissitude. In: AD, vol. 78, Wiley, London (2008)
[6] Greve, J., Schnabel, A. (eds.): Emergenz. Zur Analyse und Erklärung komplexer Strukturen. Suhrkamp, Berlin (2011)
[7] Lally, S., Young, J. (eds.): Soft Space. From a Representation of Form to a Simulation of Space. Routledge, New York (2007)
[8] Somol, R., Whiting, S.: Notes around the Doppler Effect and Other Moods of Modernism. Perspecta 33, 72–77 (2002)

Design Environments for Material Performance

Martin Tamke, Mark Burry, Phil Ayres, Jane Burry,
and Mette Ramsgaard Thomsen

1 Designing for Material Performance

The research project that induced the Dermoid (Fig.1) installation investigates the making of digital tools by which architects and engineers can work intelligently with material performance. Working with wood as a material, we were especially interested how the bend and flex of wood, can become an active parameter in the digital design process. Traditional building structures facilitate load bearing through a correlation of compressive and tensile forces passing loads linearly through the building envelope. However, materials hold internal forces that can be incorporated into structural systems thereby reducing material use and leading to a more intelligent and potentially sustainable building practice. [2] [6]

The project addresses one of the key problems in designing for material performance which lies with the traditional understanding of architectural design space. Here a linear process is prescribed from thinking the overall to the detailed and making. A material practice is yet interlinked and inherently complex. [8]

2 A Material Focus in Collaboration and Project Scope

The research project employed a practice based research method focusing on the development of speculative models and working prototypes. This material evidence provides a basis by which the customized digital prototyping tools could be validated against the limits of working with material behaviour (Fig.2).The project inquired answers in a series of empirical studies undertaken through workshops

Martin Tamke · Phil Ayres · Mette Ramsgaard Thomsen
Centre for IT and Architecture (CITA), Royal Academy of Fine Arts Copenhagen, Denmark

Mark Burry · Jane Burry
SIAL, RMIT Melbourne, Australia

Fig. 1 Dermoid at the 1:1 exhibition at the Royal Academy of Fine Arts / Copenhagen

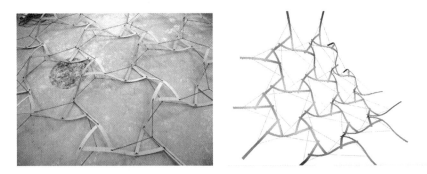

Fig. 2 Workshop summer 2010: material prototype of a reciprocal system and digital emulation

with the core project team as well as students from the School of Architecture. In-between the workshops members of the core-team advanced tool and material level with prototypes and material samples and prepared the next common event.

The work soon centred on the question how we could design with larger aggregations of distributed structural members on more complex surface topologies than a dome. We explored the material potential of concepts such as self-bracing, pre-stressing and reciprocal load bearing systems. Integrating reciprocity was not only challenging the standard computational set of tools but questioned as well the traditional understanding of reciprocal structures. As these usually utilize solely gravity [3] they ideally have to be horizontally oriented. Our interest was, however, to achieve spatial solutions that can form enclosure. Workshop results showed that the reciprocal elements could very well be oriented vertically utilizing a joint system where three beams connect in a circular manner. The definition of reciprocal structures was as well widened when we introduced a secondary stabilizing level to accommodate the vertical components of our structures (Fig.3).

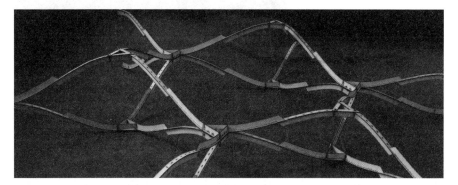

Fig. 3 A single hexagonal element of the final structure

3 Developing the Digital Modelling System

Thinking of a digital design environment with seamless interfaces as in the Digital Chain [4] a initial analysis of our digital process made us divide investigations into basic focus areas.

3.1 Space of Complexity

The first level dealt with the definition of the space that could complement a basic program as inside, outside and entrance. As a continuous topology was required for further steps an ellipsoid was ideal as the variable curvature on all three axes guarantees that each structural element will be unique. Later inquiries continued into non quadric surfaces that could follow program through freeform bodies and splitting (Fig.4).

3.2 Pattern Generator

The second level of tool development dealt with the generation of structural patterns to inhabit the initial shape. In comparison to non-repetitive patterns a hexagonal pattern suited the nodal requirements of the reciprocal system better. Working with

Fig. 4 Speculative model via a reciprocal system shaping a manifold surface

the abstract representation of axis and nodes in the pattern allowed us to reduce the complexity in this phase.

3.3 Wrap and Projection

The generated pattern is afterwards distributed on the initial shape. We investigated different approaches as projecting, mapping and growing the pattern on the topology. Contemporary methods which compute patterns in two dimensions and wrap these onto a shape distorted unfortunately so much that the patterns original properties are lost: the elements become too long or too short. More successful techniques were those that facilitate multi-step projection methods [10] (Fig.5). The best results were accomplished with relaxation or approaches utilizing self organisation (Fig.6). These allowed us to incorporate material properties as geometrical constraints as length, tolerances or trusses with single curviness. The embedding of these functions within the design environment allowed us to design lateral in a network of dependent functional, tectonic and aesthetic considerations.

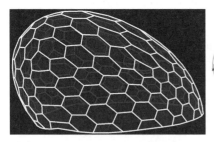

Fig. 5 Pattern achieved on ellipsoid test body using polyhedron propagation

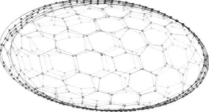

Fig. 6 Harmonising the average length and angle of the projected members using dynamic relaxation

3.4 Inhabitation

The patterns points are the nodes for the structures beam level (Fig.7). Here material parameters defined the shape of the structural elements and took into consideration meeting angles, offsets and joint details. In order to maintain a reasonable computation performance the representation happens by curves.

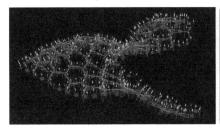

Fig. 7 A prototype populated with the materially aware beam level

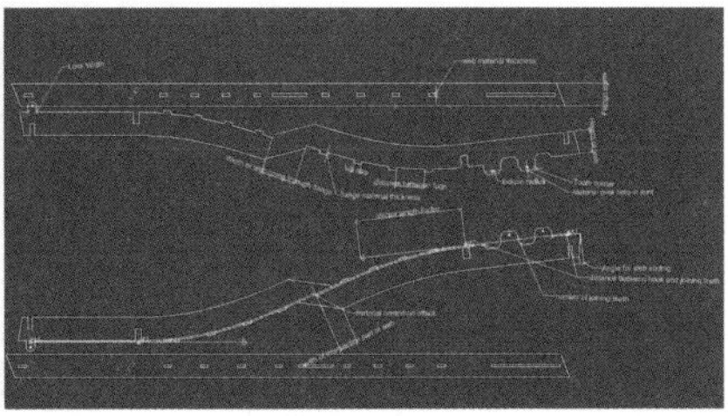

Fig. 8 2D fabrication data with material and detail related parameters

3.5 Production

Finally 2D fabrication data is extracted from the three-dimensional representation (Fig.8). This had to happen in a reasonable time as all elements are bespoke (Fig.9). In order to eliminate heavy 3D-solid calculations all material and production related geometry as spatial intersections, material thickness and offsets are computed in 2D-space. The extensive prototyping on actual fabrication machines generated the crucial feedback from material for the system.

Finally the digital toolbox consisted of a mash-up of Rhino with Grasshopper™, Digital Project™, Processing™, Open Cascade™ and Maya™. Yet parameters

Fig. 9 Detailed view of the bespoke elements

from material behaviour were introduced only in specific steps without overall cohesion. In contrast our material studies showed that the specification and inherited performance of the bend material directly influenced the overall system behaviour. Finite element simulations of parts of the structure with a detail level down to the plywood layering revealed that it is principally possible to simulate the overall structural behaviour of the Dermoid. Yet the entailed time and labour and especially the rigour on design level prohibited this approach from being used in a reasonably timed design process. In other words, what can be computed using todays technology is insufficiently rapid to offer anything like the real time feedback that such projects will ultimately derive benefit. Despite this material behaviour on one level usually affects more than the previous step. The establishments of a bottom-up logic through loops between different tools proved hence difficult and inflexible.

4 Solving Multiple Constraints Simultaneously

An alternative approach was to integrate parameters from different scalar levels and their mutual dependency in a digital environment that was to negotiate several layers of constraints. Based on the experiences from former projects [1] we looked at the physics solver incorporated in the nucleus engine of Maya^{TM}. We were able to integrate the first three steps of our design process using polygons and Mayas^{TM}constraint system. Modelling with polygons solely was flexible and lightweight. A interface to the subsequent production system in Rhino allowed to check whether the generated node system was viable for making.

Although the engine is not at all based on real world physics it gave results that could be validated in our physical models. Its ability to negotiate multiple constraints seemingly in parallel allows us to achieve the contrary targets of the structure in respect to equal distribution of cell sizes, equal angles between members in joints and the need for tectonically stable formations. Here the introduction of an inner overpressure resulted in ark and dome like conditions that resemble the initial idea of the dome.

For the nucleus engines time-based process the hexagonal topology was developed first (Fig.10). The linked polygons were subsequently constrained to boundary conditions as the points to meet the ground, the various heights and any opening. Based on our material test each vertice is furthermore informed about its maximum length, twist or bend angle in relation to other elements. A process was established that can be best described as "inverted relaxation". Here it is not a target shape on which elements positions are optimized but rather a form finding process in which inner relations are given and the design takes place in the adjustment of boundary conditions.

This was especially important as we found that the change of inner parameters as the amount of cell sizes helped little in addressing the design goals. It was rather the ability to change the polygon topology rapidly while maintaining the structural and materials boundaries that allowed us to iterate different spatial configurations.

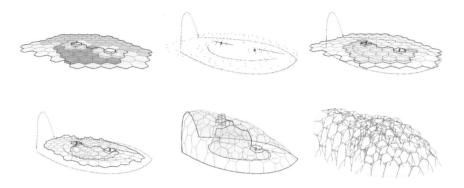

Fig. 10 Development of the structure using a time-based process

5 Production and Exhibition

The demonstrator Dermoid was designed, optimized, fabricated and assembled in less than two weeks in the period from Monday 28/02-2011 to Friday 12/03-2011 by a team consisting of 3-6 people at any given time (Fig.11). The comparison of a laserscan of the installation and the design environments calculated geometry revealed their qualitative correlation (Fig.12).

Fig. 11 Assembly of th.63e structure with temporary suspension

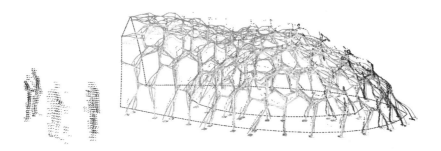

Fig. 12 Overlay of computed geometry (lines) and laserscan of relaxed (grey dots) and tensioned state (black dots)

6 Conclusion

The approach taken here (Fig.13) facilitates parallel investigations in the best use of material and the closest match to predefined design criteria. The desired real-time computed feedback of the precise performance for complex structural arrangements could yet not be achieved. We were however able to establish a design system that integrated material behaviour to a degree that allowed us to sufficiently predict a structures overall performance. We were able to proceed with ready information in a design space with simultaneous contributions from architects, engineers, material and computer scientists.

Fig. 13 Finalised Dermoid

We found that the technical answer for a question of our complexity does not lie in the exact replication of material behaviour through simulation but in the overlay of a multitude of rough internal conditions. The possibility to return these to geometrical rules was most purposeful for our structural suit. And the combination of empirical testing and dynamic input into our bespoke digital design environment gave us a very different result than we would have arrived at if we tried to solve the problem through optimisation alone.

Here it is a dialogue between the numerically captured parameters from material and structure and the design intent. Both the designer and the generative tool hold now the ability and obligation to adapt and change on the close to realtime feedback they receive. We found though that even if the generative environment bears this ability we could not easily reach the design goals by changing the inner parameters of the system [11], as here the resolution or shape of elements. The changes had to extend to the outer - the framing parameters. These were in our case mainly those connected to topology.

This observation challenges the promise of contemporary generative techniques [9] to solve more complex questions of architecture. It approves the observation of other projects [7] were a similar linkage of multiple parameters is ultimately narrowing the space of design to an isolate solution. The result might yet not fulfill all criteria in a sufficient manner.

Yet the subsequent extension of the solution space excels the frame of the generative system and questions the protocols of our design environment. Were contemporary parametric and CAD tools inherit the legacy of traditional linear work processes Dermoid asks for fluent and extendable environments. These offer alternative means to organise and discuss design parameters and can ultimately keep up with the speculative nature of design.

Acknowledgements. Dermoid is supported by the VELUX Visiting Professor Programme 2009- 2010 of the Villum Foundation. Photos: Anders Ingvartsen. Project team: Mark Burry, Mette Ramsgaard Thomsen, Martin Tamke, Phil Ayres, Jane Burry, Alexander Pena, Daniel Davis, Anders Holden Deleuran, Stig Anton Nielsen, Aaron Fidjeland, Morten Winther, Tore Banke, Michael Wilson and students from the departments 2 and 10 (Copenhagen).

References

1. Deleuran, A., Tamke, M., Ramsgard Thomsen, M.: Designing with Deformation - Sketching material and aggregate behaviour of actively deforming structures. In: Proceeding of Symposium on Simulation for Architecture and Urban Design, SimAUD 2011, Boston, MA, USA, April 4-7 (2011)
2. Hensel, M., Menges, A., Weinstock, M.: Emergent technologies and design: towards a biological paradigm for architecture. Routledge, New York (2010)
3. Popovich Larsen, O.: Reciprocal Frame Architecture. Architectural Press, London (2008)
4. Schoch, O.: Applying a digital chain in teaching CAAD and CAAM. Swiss Federal Institute of Technology, Faculty of Architecture, Zurich (2005)

5. Simon, H.A.: The Sciences of the Artificial, 3rd edn. The MIT Press, London (1996)
6. Schwitter, C.: Engineering Complexity: Performance-based Design in use. In: Kolarevic, B., Malkawi, A.M. (eds.) Performative Architecture Beyond Instrumentality. Spon Press (2005)
7. Tamke, M., Ramsgaard Thomsen, M.: Implementing digital crafting: developing: its a SMALL world. In: Proceedings of the conference Design Modelling Symposium 2009, Berlin, pp. 321–329 (2009)
8. Tamke, M., Riiber, J., Jungjohann, H.: Generated Lamella. In: Proceedings of the 30th Annual Conference of the Association for Computer Aided Design in Architecture, ACADIA (2010)
9. Terzidis, K.: Algorithmic architecture. Architectural Press, Oxford (2006)
10. Wester, T.: The structural morphology of basic Polyhedra. In: Francois Grabiel, J. (ed.) Beyond the cube. Side, pp. 301–343. John Wiley and Sons, Chichester (1997)
11. Wilson, E.O.: Sociobiology: the New Synthesis. Belknap Press, Cambridge (1975)

Faserstrom Pavilion: Charm of the Suboptimal

Mathis Baumann, Clemens Klein, Thomas Pearce, and Leo Stuckardt

Abstract. Current discourse on form finding, material systems and performance tends to regard optimization, control and digital manufacturing as logical consequences of computational design. This paper challenges such assumptions by presenting a pavilion, recently built as a student project at the Technical University of Berlin. While using parametric modeling tools, contextual constraints induced a low-tech and hand-on production, using ready-made materials and incorporating standardized details. The pavilions expressive form was conceived in an early stage and the suboptimal was appropriated as an implication of the further design process.

Fig. 1 View of the pavilion from the waterside

Mathis Baumann · Clemens Klein · Thomas Pearce · Leo Stuckardt
TU Berlin, Faculty of Architecture

1 Introduction

1.1 Discourse: Control, Optimization, File to Factory

In recent years, architects and designers have shown a growing interest in the so-called "performative potential of wood". Material deformation becomes a design strategy rather than being avoided, instrumentalizing woods inherent characteristics such as its anisotropic and heterogeneous nature. The proliferation of parametric modeling software has opened up before unknown possibilities to integrate the simulation of and control over material behavior into the design of intelligent material systems. Overcoming design as a consecutive process, a more integral approach is suggested, defining form finding as a digitalized iterative process and allowing for analytical feedback into generative algorithms (Menges 2010).

In the discourse on material systems, some developments can be critically observed. Mostly developed as an exercise in an academic context, a scientific approach aiming at exact simulation and optimization tends to overshadow more classical design parameters such as context or atmosphere. Furthermore, as control becomes such an important criterion, "file to factory"-discourse suggests the use of computer aided manufacturing as a logical consequence of this integral approach to design: rearranging the classical workflow, the architect is put in charge of an (alleged) seamless chain of design to production. Directly processing the output of computational algorithms, advanced (scanning, milling, cutting) technology is employed to transform materials to the use of material systems. Mass customization replaces standardization.

1.2 A Critical Stance: Contextual Performance

The Faserstrom pavilion, whilst utilizing parametric modeling software, may present a critical stance in this context as it relativizes the imperative of the described categories - control, optimization, CAM. The pavilion was built in the spring of 2010 by the authors, four architecture students at the Technical University of Berlin (Prof. Regine Leibinger). It was developed during a one-semester design seminar aimed at creating innovative wooden structures in parks surrounding manor houses on the Isle of Rügen.

Fig. 1.1 Model studies Scale 1:20

Interaction with this context constituted the driving force behind the design of the pavilion. This applies to its gestalt as well as to its constructive system. The pavilions filigree structure, a network of fine trusses composed of bent wooden slats, was conceived as a mimetic extension and an overhead bundling of the reeds surrounding the secluded peninsula Liddow. The formal principle of this structure emerged from early and intuitively built models and was but slightly modified. Throughout the design process, these "prototypes" were regarded as a reference, guiding the design decisions and their application in the parametric model.

Likewise, the constructive logic of the pavilion emerged from contextconstraints: the site's remoteness and lack of technical infrastructure, a very tight budget, limited man-power and an extremely short time scheduled for production and assemblage. These factors induced the choice for a hands-on, low-tech and cheap construction method. Every detail was designed to be carried out on site by the authors using only basic portable electrical tools. Joints and footings were standardized and a ready-made chosen as the basic material.

Fig. 1.2 Standardized footings and joints

1.3 Material System: Beyond the Gridshell

Through bending and joining pine slats, commonly used as skirting boards (5 x 60 mm), are transformed from a ready-made into a dynamically bent spatial sculpture. Trusses are woven into a network, in which every added truss braces the overall structure. The segments composing these trusses each consist of two bent slats that form a reciprocal bending-active equilibrium.

Whereas initial experiments pointed into the direction of a gridshell, the subsequently developed material system offered both technical and spatial advantages to this precedent. As each segment and thus each truss can be regarded as rigid, a lightweight punctual foundation appropriate to the pavilion's temporary nature replaced the gridshell´s heavy clamping at the foundations. Secondly, whereas the topology of a gridshell can be described as a single surface, a shell with a distinct inner and outer side, the Faserstrom pavilion constitutes an additive woven network, a semi-lattice continuing the texture of the surrounding reeds and mirroring its underground rhizomatic structure .

Nevertheless, the material system shares one important advantage with the gridshell: standardization. One module, a slat varying in length but constant in section, could be used for the entire structure. Joints, nodes and footing were also standardized. As both elements of one segment have the same tangents at their joints and the nodes between trusses are always horizontal, all connections could be made using simple bolts, thereby avoiding the need for custom-made wedges. All footings are vertical so that they could be built using standard dowels out of cheap construction wood.

Fig. 1.3. Section

2 Modelling Process

Parametric software (Grasshopper for Rhino) was an indispensable tool to design and build the pavilion. Firstly it enabled the control over the material system by tracing it´s components and automating their generation. Secondly it integrated the standardized details necessary for the DIY-character of the building process.

Fig. 2.1 a Parametric model   **Fig. 2.1.b** Manual production of a truss

2.1 Segment Algorithm: Intuitive Simulation

While other elements were standardized, the form of each truss segment, the length and curvature of its elements were variable. A three-dimensional database of empirical values was created, out of which each individual segment could be parametrically interpolated (Fig. 2.2). In a first step, individual segments were built, photographed and traced. Multiple curvature definitions available in Grasshopper were tested to trace the curvature of the shorter (called primary) and longer (called secondary) elements of one segment. The best approximation was reached through Bezier curves, which had their starting and end point in the apex and vertex (joint) of the element and were mirrored over the apex to complete the curve.

It was possible to reduce the description of one segment to a minimal combination of three factors: (1) the length of the primary curve, (2) the angle of the vertex tangent, (3) the position of the apex of the secondary curve in relation to the vertexes, described as the distance between the apex and the center of both vertices. All tangent lengths needed to further describe the Bezier curves, could be described as functions of one of these three factors.

In a next step, a large variety of segments was traced and for each of them, these three factors were defined. Subsequently a 3-dimensional database was created by understanding each of the three factors as respectively the x, y and z coordinates of single points in Cartesian space. Each point thus describes the specific geometry of one truss segment. By interpolating these points into a surface, a 3-dimensional database for the range of all possible curves was established. Conversely, for any given primary curve (described by its length and tangent angles) the information to define its respective secondary curve could be retrieved from this database.

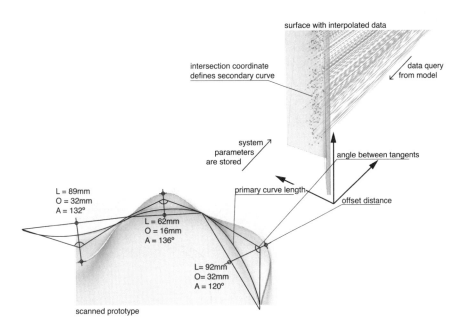

Fig. 2.2 3-dimensional database based on the relevant parameters of traced trusses

Though rather intuitive, this strategy proved to be successful. The segments that were simulated through interpolation showed to be, when built, a fairly exact approximation of their digital model. The 3-dimensional database could account for the mutual interdependency of the curvature of both elements in one segment – which could not have been described or simulated using conventional material parameters such as elastic modulus or minimal bending radius. Rather than material parameters one could speak of system parameters that are specific to the developed material system.

2.2 Pavilion Algorithm: Line to Network

The modelling process consists of two phases (Fig 2.3). In a first phase, an algorithm creates the overall geometry of the primary curves describing the general shape of the pavilion. As the trusses are oriented in a vertical plane, their density and position can be controlled by drawing lines (the trusses´ projection) in plan. Their intersections that are to represent actual nodes are manually selected, forming a woven network and respecting a minimal distance between the nodes.

The Grasshopper definition starts by projecting the selected nodes vertically onto a surface controlling their height. Then, Bezier curves are created between endpoints and nodes and between nodes of every single truss. Standardization is integrated by defining the tangents of the Bezier curves as vertical at the endpoints and horizontal at the nodes. As a truss must have at least two nodes to form the truss network and these nodes are made horizontal, every truss inevitably obtains an inflection points in the Bezier curve between these nodes.

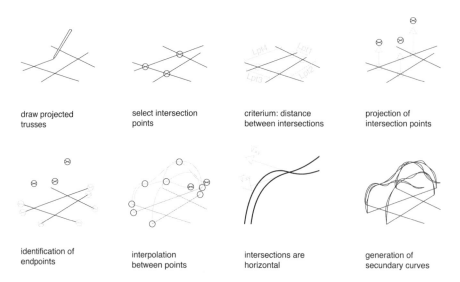

Fig. 2.3 The entire modelling algorithm of the Faserstrom pavilion

Fig. 2.4.a Erection of the trusses at the site   **Fig. 2.4.b** Stored trusses

In the second phase, these curves are divided into smaller primary curves and secondary curves, necessary to achieve these primary curves´ shape, are generated by using the segment algorithm described above. Since the customization of the individual segments is only manifested in variable slat lengths, the algorithm produces a simple numerical output instead of complex cutting patterns. With an Excel list and the ready-made materials, the pavilion could be produced and erected "from scratch" on site by seven students in less than one week.

3 Evaluation

3.1 Simulation and Optimization

Analyzing the completed built structure, substantial differences could be observed when compared to its digital model. The upper section of each truss sagged, whereas the lower (more vertical) parts bulged out. This in turn influenced the curvature of the individual segments and rendered them incongruent to their digital parent. Though the behavior of isolated single members was successfully simulated and its values were fed into the form finding algorithm, the digital model of the overall structure was based on geometrical rules rather than on physical simulation. Implementing a physics solver to simulate the "aggregate behavior" of the structure during the modeling process would have become highly complex given the interdependency of all elements in the material system (cf. Deleuran et al. 2011).

If the Faserstrom pavilion is considered a prototype to be further developed, its material system could certainly be optimized in many ways. To name but a few, its joints, the trusses´ weak points, could be further reinforced. By adding branching truss segments or rethinking the horizontal nodes, the system-inherent inflection points could be compensated or avoided. Replacing vertical by inclined footings would optimize the load transfer.

3.2 Conclusion: Charm of the Suboptimal

Returning to Liddow two months after the building workshop, we found the pavilion not broken but sagged to the ground, resembling a giant bird's nest rather than our computer model. A good many of visitors preferred this accidental to the planned form. Whether or not sharing this opinion from an aesthetic point of view, it was a good lesson for us as the designers of this structure. Of course, a physics simulation could have warned us and yes, optimization of the material system could have (partly) prevented this.

But if a physics simulation would have gone beyond the scope, time-budget and technical abilities of the design team, only few of the optimization suggestions as described above would have been compatible with its design intentions. Inclined footing for example would have reduced the mimetic blending into the surrounding reeds. Forming trusses as arches would have disabled the network character of the overhead structure. While increasing structural performance these adaptations would have reduced spatial complexity and contextual performance.

As a specific answer to its context rather than an optimizable prototype, the Faserstrom pavilion had nothing to do with ideal or hierarchical structures "of such a kind that nothing could be added or taken away or altered without making it less pleasing" or stable. Stubbornly (all too stubbornly?) following early ideas created in a short phase of actual form finding the suboptimal and redundant were accepted as charming byproducts of design.

Looking beyond failure and success of the pavilion as its final product, the approach presented here takes a challenging stance in the current discussion on digital design and production. It shows that advanced computer-aided planning does not compulsorily have to lead to high-tech robotized production. In this specific case, the opposite assumption has lead to the transformation of a ready-made material into a complex and innovative structure, which atmospherically and technically blends into its surroundings.

Acknowledgments
Chair for Building Construction and Design (TU Berlin), Prof. Regine Leibinger
Assistant Professors:
  Matthias von Ballestrem
  Cornelius Nailis
  Martin Schmitt

Chair for Design and Solid Construction (TU Berlin), Prof. Mike Schlaich
  Assistant Professors:
  Annette Bögle
  Christian Hartz

References

Deleuran, A.H., Tamke, M., Ramsgard Thomsen, M.: Designing with Deformation – Sketching material and aggregate behavior of actively deforming structures. In: SimAUD 2011 (2011)

Menges, A.: Material Information: Integrating Material Characteristics and Behavior in Computational Design for Performative Wood Construction. In: Acadia 2010 (2010)

Rhizome - Parametric Design Inspired by Root Based Linking Structures

Reiner Beelitz, Julius Blencke, Stefan Liczkowski, and Andreas Woyke

Abstract. The project is based on the development of a programmed growing process found in the example of the reed's root structure, the rhizome. This paper shows the projects development including the experimental research of basic structural qualities, the development of digital, generative design-tools and the realization of the generated structure on site. Furthermore the question about the designer's degree of influence on the growing process without loosing its randomness arouses great interest in us. Therefore the project examines a concept which introduces specific spatial presets to intuitively control the generation of the growing algorithm into a usable structure.

Fig. 1 View from the Manor House

Reiner Beelitz · Julius Blencke · Stefan Liczkowski · Andreas Woyke
Technical University Berlin - Faculty of Architecture

1 Introduction

In the winter semester 2009/ 2010 we took part in a student competition, commissioned by the TU Berlin aiming at the design of a temporary wooden structure using parametric software. The workshop, named „ *Prototypisches Entwerfen"*, started off with the visit of several sites on the island of Rügen. We had to choose one specific site and come up with some first rough ideas. The presence of the lake´s water, spacious fields, trees and the belt of reed impressed us instantly. All the natural qualities were easy to catch but nonetheless we felt that something was missing: the physical experience of the reed and associated with that the connection of the elements land, shoreline and water. So we set our focus on the analyses of the reed, its behavior, special qualities and its natural reproduction. During that process we came up with the idea to imitate its natural pattern of growth to generate a mimetic structure creating a specific interaction between the structure and the reed and thus creating a space.

1.1 Analogue Growth Studies

To understand the nature of growth we analyzed possibilities of natural reproduction and different orders of linking systems. Translating the reproductive qualities of the reed`s roots system to create a spatial structure which is highly cross linked and permeable was our main task. Based on various experiments we defined a node system with a set of two node plates and clamped laths in between. To verify the rigidly of the primary structure we built a first full-scale mockup. Furthermore we abstracted and systemized the rhizomatic growth strategies to constitute a digital model.

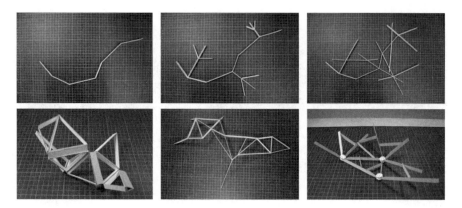

Fig. 1.1 Analogue Studies

1.2 Adaption of Growing Rules

The adaption of a natural growth led to a range of necessary abstractions. The roots were translated into lines and points representing the nodes and internodes of the reed. According to the first structural models we started to simplify the basic growing principles of the reed into a logical sequence.

Fig. 1.2 Basic Growing Rules

At the beginning of the growing process a set of points is placed by the designer (1). The points develop randomly oriented branches and create a second generation of points at the end of the edges (2). Depending on a field of parameters the points including their corresponding branches can be deleted or remain in the structure as the second generation (3). The basic process is repeated with following generations for several times. Additional rules are introduced to control the process of growth and meet constructive requirements. Between the edges only minimal angles are allowed (4/5). The points are enabled to create new links to other points (generation-spanning) (6).

2 Development of Digital Tools

2.1 Introduction

Based on the previous studies of root systems and the decision to generate growth it was necessary to find a tool that could transfer the scheme of rhizomatic reproduction into a generative process as well as to generate a geometry. The 3D-modelling software *Rhinoceros* supplied those qualities. By creating a specific "growing simulation" code, *Rhino Script* delivered a topological structure in the form of a "point-line cloud". Afterwards the given geometry was shaped using the plug-in *Grasshopper*.

2.2 Grasshopper Model

A parametric Grasshopper model generated the final geometry of nodes and laths based on the point-line-cloud created by RhinoScript. The generated input of the lines was translated into laths, the vertices into nodes and the corresponding points that constitute the normal orientation. The node was designed by using the projection of adjoined lines on the plane of the node and their angle bisectors. The result was a parametric model whose components could be altered regarding the dimensions of the details.

2.3 Development of Growing Algorithm

The implication of the growing process into code presented itself as a repetitive process. First examples were produced and results were analyzed regarding structural, intuitive and aesthetic aspects. Afterwards certain qualities and problems were extracted and new optimized rules were implemented into the script.

Fig. 2.1 First Results of Growing Sequences

The use and alteration of single basic rules of the script such as "number of branches", "length of branches", and "connectivity of existing points through branches" changed the characteristics of the outcome of the script drastically. Nonetheless the necessity of influencing the growth furthermore to constitute a usable structure became clear. We need to implement more intuitive design rules to control such a growing process but without loosing the element of "randomness" and natural appearance. The space where the grown structure can spread needs to be limited by a certain degree and additional factors.

2.4 Spatial Concept

The aim is to define potential growing areas and to limit the expansion of branches to regions. Points are being placed manually into *Rhino* and act as parameters. The growing depends on assigned distances between the points (visible as spheres) and takes place in defined boundaries There are 3 categories of points that offer different qualities: *"Hull"* aims at creating an inner space enclosed by structure , *"Path"* creates a denser area to walk on and *"Block"* disables the growth of branches within a certain radius.

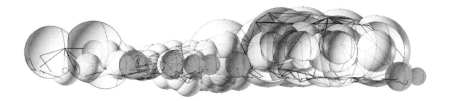

Fig. 2.2 Section of Spatial Presets and Point-Line-Cloud

The output of the modified script offered imaginative qualities and verified the success regarding the use of the generative algorithm to shape space. Nonetheless the complexity of the point-line cloud regarding the amount and density of the elements made it impossible to assign the details of the nodes without causing imprecision on the joints. A solution for the orientation of the nodes had to be found.

2.5 Concept of Node Orientation

2.5.1 Introduction

According to the spatial concept in combination with the developed detail, several structural decisions had to be made. A discrete orientation of nodes and laths was apparently required to shape the structure within the given range of angles between the components. Furthermore we expected a certain ratio of triangles to be required for the stability of the structure. Due to the expected "low tech" construction of the pavilion on site, an easy assembling of the components was particularly important. To avoid inaccuracies at the joints of lath and nodes, we estimated that torsion-free nodes would be crucial. Therefore we implemented the orientation of nodes and laths into the growing algorithm. On the one hand the tendency of the initial algorithm to produce triangles in between adjacent points was obvious. On the other hand we were aware about the impossibility to create torsion-free nodes in triangular meshes. By including the rule for torsion-free nodes into the algorithm, meshes only consisting of triangles would be impossible. We made the assumption that the supplemented algorithm will still produce a sufficient amount of triangles as well as quadrilaterals or other configurations to provide a stable construction.

2.5.2 Node Orientation

To assign an orientation to the associated node, a second point close to the node was introduced representing the orientation of the node`s normal. This information is a crucial factor for the spreading and the connectivity of points. To fulfill constructive requirements rules had to be set for the design of the node. To prevent the intersection of laths at the nodes we defined a minimal angle in between the laths related to the node`s plane. Due to stability we realized that a certain cross sectional area of the laths wood-fibers between two nodes needed to remain after the laths had been beveled. Therefore we set a minimum and maximum angle between the normal of the node and the associated lath.

2.5.3 Orientation Process

For the first points of the growing algorithm placed by the designer, the orientation is determined to the z-axis. By transferring the orientation to the second generation a plane or a torsion free lath can be constructed In between both nodes. The node orientation of the new generation can approximately be aligned to the direction of the closest parameter point (Hull) on the lath`s plane. A new point is set at the optimized orientation. Each generation of existing points tries to develop branches to adjacent points. In general there are two possibilities of connections. 1. If one of the points has only formed one connection to another point so far, the new orientation will be solved by the intersection of the new and the old edge (lath) planes. 2. If the orientation of both nodes is similar, they can connect without structural changes.

Fig. 2.3 3D Model with Discrete Node Orientation

2.5.4 Conclusion Node Orientation

Using the final Algorithm we generated several versions of structures for the site (Example Fig 2.3). The results consisted of precise oriented joints and laths and the angles between the components stayed within the permitted range. Nevertheless the results considerably varied in density, in the number of connections and in the general shape. Viable configurations appeared partly or in larger sections but not entirely.

3 Realization

For the realization on Rügen, we chose a simplified algorithm with a uniform orientation to the z-axis. Without any professional static analysis and therefore no official static verification and experience we decided to build a structure whose reliability was assessable. The uniform component orientation towards the z-axis led to the possibility to add required static elements. The laths (standard pinewood / 30mm x 140mm) were cut-to-length and trimmed in „Abbundzentrum Leipzig" based on the provided 3D model. The nodes consisting of two panels (plywood, maritime pine, 20mm) were cut out by a CNC milling-machine based on a 2D drawing of grasshopper. Afterwards the components were delivered to Rügen. The tagged nodes and laths were sorted on site via 3D-model and plan which was supplied by "Abbundzentrum Leipzig". Subsequently the elements were assembled in manageable units, the site was measured and the foundations were set successively regarding to their units. The parts were placed step by step and assembled to the structure on site.

Fig. 3.1 Section of Realized Structure

Fig. 3.2 View of Realized Structure

4 Conclusion

Considering the development of the project, the built structure constitutes a part of our experimental work on growing algorithms. If we consider all the circumstances the realized structure was the feasible project at the given time. The process, influenced by the development of digital design tools and the exploring of its spatial possibilities, was iterative and open-ended. The concept of limiting the space to grow is an approach to find a balance between a defined spatial perception and the natural appearance of a grown structure. In general we were successful in influencing the shape of the structure to a certain degree. It developed intended situations but the generated structures were unpredictable in terms of their stability and feasibility. At this point the generated results, without any modifications, are rather sculptures than safely usable structures. With regards to a further improvement of the algorithm, a more variable handling of the generated parts during the growing process would be conceivable. Ongoing tests of the nodes connectivity and a rating of fitness could result in the adjustment of ineffective parts. Single branches and points could be rotated or moved as well as deleted to optimize or facilitate new connections. This could result in a more adaptable growing process in terms of connectivity and stability.

Kinetic Pavilion
Extendible and Adaptable Architecture

Corneel Cannaerts

Abstract. This case study describes Kinetic Pavilion a master student project by Elise Elsacker and Yannick Bontinckx developed for elective Parametric Design & Digital Fabrication at the MMLAB of Sint-Lucas School of architecture. The pavilion represents a building that can adapt its formal layout and shape to different uses and input parameters. The Kinetic Pavilion is designed as an open platform that allows different modes of adaptation and interfacing with different open source technologies. The paper looks at the relation between parametric modeling and adaptable, kinetic architecture and its implications on the role of the designer and the notion of the model in architectural design.

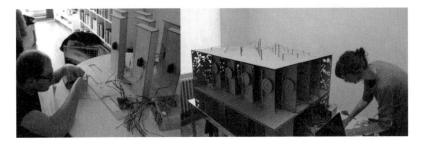

Fig. 1 Kinetic Pavilion model

1 Parametric Model as Exploration Machine

A significant implication of introducing parametric modeling[1] in architectural design is the shift from a linear to a non-linear exploration of variations in design. In

---

Corneel Cannaerts
Sint-Lucas Architectuur, MMLAB, Gent, Belgium
RMIT University; SIAL, Melbourne, Australia

a linear design modeling process the model is a gradual accumulation of design decisions. Exploring variations often requires backtracking in the design history and branching in a different direction. The designer works directly on the geometry representing his design ideas.

Parametric modeling introduces a dual representation: on the one hand a system is constructed by explicitly defining relations between parameters, on the other hand geometric results of this system are examined. The designer influences geometry in an indirect way by altering input parameters or changing the rules of the parametric system.

As a consequence the model becomes a machine for simultaneously defining a solution space and exploring variations within this solution space. The focus of the designer shifts from designing objects -represented by geometry - to designing systems – represented by a parametric definition, script or algorithm.

2 Translating from Model to Building

A crucial part of any design modeling process is anticipating the translation between model and the artifact or building it represents. Advances in digital fabrication allow for a precisely controlled machining of building components that incorporate the variance made possible by parametric systems. While the digital reaches out into the physical, material properties, fabrication and assemblage constraints enter the digital model.[2] Parametric modeling facilitates the translation between modeling and building.

Fig. 2 Laser-cut gears

3 Building Machines

As outlined above parametric modeling provides an excellent tool for design exploration and can be constrained in such a way that it will result in viable material

---

[1] In this paper parametric modelling is defined as any process where in parameters are used to control geometry, so it can refer to scripting, programming or patching.

[2] What Axel Kilian calls 'bidirectional constrains' or Neri Oxman refers to as 'factory-to-file'.

solutions. Nevertheless actually constructing a building or artifact from a dynamic parametric model is often a reduction. It freezes the rich and varied set of possible design solutions (all fitting within the material constraints set-out by the fabrication and assemblage process), to a single fixed design solution that fails to communicate this richness. In some cases the richness of the parametric is successfully communicated on component level, in other cases this reduction is necessary and actually the reason for the parametric model to built (e.g. in optimization an optimal solution is negotiated and build).

An alternative to possible this reduction of a design system into a fixed and final solution is actually building the system rather than an object derived from it. Thinking of architecture as a machine or a system is of course not a novel idea, many aspects of architectural design have used the machine either as metaphor or as inspiration.[3]

Fig. 3 Arduino controller and wiring

4 Modes of Adaptation

The Kinetic Pavilion was a response to Parametric Design & Digital Fabrication elective, framing the questions outlined above. After a general introduction in the subject matter and an introduction to working with Grasshopper, students were asked to look specifically at ways in which Grasshopper could be adapted and extended through external software and hardware.

The Kinetic Pavilion was designed in Grasshopper and fabricated using a lasercutter at MMLAB. The scale model of the pavilion was set up as a first step in testing necessary hardware and software configurations, try out different modes of interaction, and iterate through design ideas. Building on open source hardware, software and digital fabrication techniques it tries to extend the toolbox architectural design. The model itself is developed as an open and extendible structure that can respond to variety of input data.

The pavilion is consists of 28 columns, arranged in a triangular grid and connected at the top with linear bearings and tensile structures. The height of the

---

[3] See Nicolas Negroponte: Soft Architecture Machines.

columns can be altered by a servo motor and laser-cut gears connected to an Arduino micro-controller. Values for the heights where mapped and communicated from Grasshopper to Arduino using the FireFly plug-in.

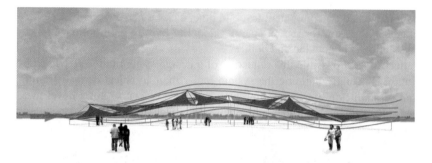

Fig. 4 Ventilation

Weather conditions: In cold climates the surface of the roof orients itself in order to maximize solar gains, difference between neighboring columns can be used to control the amount of sunlight filtering through the roof. In warm climates the surface of the roof maximizes ventilation and the shaded area under the roof. In the scale model this was achieved by using Geco a plug-in that allows communication between Grasshopper and Ecotect: a mesh representing the environment and weather files for a chosen location are used to calculate solar gains on the pavilions surface.

Fig. 5 Night Render

Crowd movements: The pavilion can adapt to movement of people beneath and around the roof. This can be done accumulative, learning frequent used paths over time and reinforcing them or interrupting when events take place, or instantaneously making the pavilion a reactive sculpture. This was simulated in the scale model by tracking the movement of fingers on an iPad, communicating to Grasshopper over OSC. Tests done using a webcam and Microsoft Kinect to track movements, showed the extendibility of the system.

Kinetic Pavilion 339

Fig. 6-7 Textile Structures

Data visualization: In a more abstract way the pavilion can also be used to visualize data - by filtering out feeds, or communicating with Processing using GHowl plug-in for Grasshopper. An example sketch in processing was produced and a sketch reacting to twitter feeds. Although this showed some potential it was not worked out during the time of the course. Building and designing kinetic architecture requires architects to rethink the subject matter but also their tools. The Kinetic Pavilion benefited from open source hardware and software and would have been hard to conceive let alone build without these.

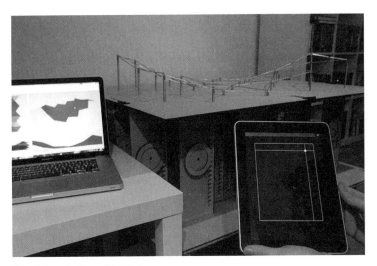

Fig. 8 Interfacing with iPad

5 Conclusion

What makes the Kinetic Pavilion interesting from a design modeling perspective, is not so much its role as a model of future building but rather in the potential in shifting from object to system thinking. As such the model of the pavilion is not a *scale model* – an object representing a scaled down version of the geometry of a building – but a *prototype* - an object showcasing the working of a system.[4]

Further research could be done and improvements could be made for the types of movement – instead of 28 actuators showing linear motion, combination between actuators could be examined - and the control system – distributed, networked controls instead of one central computer.

The scale model of the pavilion is presented in a lab context and does only handle one mode of adaptation at the time and this adaptation is *reactive* rather than *interactive*.[5] Translating it to 1:1 scale and placing it in public space would raise question as to how larger crowds would interact with it and how conflicting data inputs could be dealt with.

The Kinetic Pavilion shows the possibilities of open source software and hardware as a way of and opening up and appropriation of technology by architecture students. As such it has (at least for the students involved) extended the toolbox of architectural design.

Acknowledgments. The pavilion is a student project by Yannick Bontinckx and Elise Elsacker developed under guidance of the author and Tiemen Schotsaert.

Software used:
- Grasshopper: (http://www.grasshopper3d.com)
- Processing: (http://processing.org)
- Arduino: (http://www.arduino.cc)
- Geco for Grasshopper: [u t o] (http://utos.blogspot.com)
- Ecotect (http://www.autodesk.com/)
- gHowl: Luis Fraguada (http://www.grasshopper3d.com/group/ghowl)
- FireFly: (http://www.fireflyexperiments.com)
- Kinect & processing: (http://www.shiffman.net)

More info: www.kineticpavilion.com

References

Burry, M.: Models, Prototypes and Archetypes. In: Burry, M., Ostwald, M., Downton, P., Mina, A. (eds.) Homo Faber: Modelling, Identity and the Post Digital. Archadia Press, Sydney (2010)

Haque, U.: Distinguishing Concepts: Lexicons of Interactive Art and Architecture. In: AD Special Issue: 4dsocial: interactive design environments, Wiley, London (2007)

Kilian, A.: Design Exploration through Bidirectional Modeling of Constraints. Phd. Massachusetts Institute of Technology (2004)

Negroponte, N.: Soft Architecture Machines. The MIT Press, Cambridge (1975)

Oxman, N.: Digital Craft: Fabrication-Based Design in the Age of Digital Production (2007), http://www.materialecology.com

---

[4] See Mark Burry: Models, Prototypes & Archetypes.
[5] See Usman Haque: Distinguishing Concepts: Lexicons of Interactive Art and Architecture.

Author Index

Ahlquist, Sean 71
Aish, Robert 1
Alexa, Marc 79
Ayres, Phil 89, 309

Baerlecken, Daniel 209
Barczik, Günter 9
Bardt, Chris 17
Baumann, Mathis 319
Beelitz, Reiner 327
Blandini, Lucio 217
Blencke, Julius 327
Bletzinger, Kai-Uwe 25
Block, Philippe 181
Brandstätter, Christoph 283
Burry, Jane 89, 309
Burry, Mark 89, 309

Cannaerts, Corneel 335
Chen, Duoli 137
Clausen, Kenn 137
Coenders, Jeroen 39

Davis, Daniel 89
del Campo, Matias 225
de Leon, Alexander Peña 89, 267
Dierichs, Karola 259
Dimcic, Milos 97
Dziedziniewicz, Michal 17

Filz, Günther H. 105
Fleischmann, Moritz 239

Gengnagel, Christoph 123
Georgiou, Odysseas 115

Huber, Jörg 171

Kirkegaard, Poul Henning 291
Klein, Clemens 319
Klein, John 89
Knippers, Jan 47, 97
Ko, Joy 17
Koppitz, Jan-Peter 249
Kotnik, Toni 145
Krieg, Oliver 259
Kristensen, Mathias K. 291
Künstler, Arne 209
Kure, Johan 137

Labelle, Guillaume 161
Labs, Oliver 9
Lachauer, Lorenz 145
Lafuente Hernández, Elisa 123
Liczkowski, Stefan 327
Ljubijankic, Manuel 171
Lordick, Daniel 9, 275

Maleczek, Rupert 153
Manegold, Martin 209
Manickam, Thiru 137
Manninger, Sandra 225
Matl, Alexander 283
Menges, Achim 71, 239, 259

Nembrini, Julien 161
Nytsch-Geusen, Christoph 171

Oberascher, Martin 283

Pauly, Mark   191
Pearce, Thomas   319
Peters, Brady   89
Pugnale, Alberto   137

Quinn, Gregory   249

Ramsgaard Thomsen, Mette   309
Raun, Christian   291
Reichert, Steffen   259
Reitz, Judith   209
Rippmann, Matthias   181
Rörig, Thilo   123
Rutzinger, Stefan   301

Samberger, Steffen   161
Schinegger, Kristina   301
Schmid, Volker   249

Schuster, Albert   217
Schwartzburg, Yuliy   191
Schwinn, Tobias   259
Sechelmann, Stefan   123
Shelden, Dennis   267
Sobejano, Enrique   55
Sobek, Werner   217
Späth, Achim Benjamin   201
Sternitzke, André   161
Stuckardt, Leo   319

Tamke, Martin   309
Thurik, Anja   249

Usto, Kemo   137

Werner, Liss C.   63
Woyke, Andreas   327

List of Contributors

Sean Ahlquist
Research Assistant
University of Stuttgart
sean.ahlquist@icd.uni-stuttgart.de

Robert Aish
Director of Software Development
Autodesk
robert.aish@autodesk.com

Marc Alexa, Dr.
Professor / Computer Graphics
TU Berlin
marc.alexa@tu-berlin.de

Phil Ayres
Assitant Professor
CITA, Royal Academy of Fine Arts
Copenhagen
phil.ayres@karch.dk

Daniel Baerlecken
Visiting Assistant Professor
Georgia Institute of Technology
daniel.baerlecken@arch.gatech.edu

Günter Barczik
Research Assistant
Brandenburg Technical University
Cottbus
gb@hmgb.net

Chris Bardt
Professor
Rhode Island School of Design
cbardt@risd.edu

Mathis Bauman
Student
TU Berlin
mat.bau116@googlemail.com

Reiner Beelitz
Student
TU Berlin
reiner_beelitz@hotmail.com

Alireza Behnejad
Board Member
Archivision Company
behnejad@archi.vision.com

Julius Blencke
Student
TU Berlin
julius_bl@web.de

Kai-Uwe Bletzinger, Dr.-Ing.
Professor
TU München
kub@bv.tum.de

Philippe Block, PhD.
Assistant Professor
ETH Zürich
block@arch.ethz.ch

Christoph Brandstätter
Architect / Owner
Brandstätter ZT GmbH
cb@brandstaetter-zt.at

Mark Burry
Professor
SIAL, RMIT Melbourne
mark.burry@rmit.edu.au

Corneel Cannaerts
Researcher / Architect
Sint-Lucas Architectuur
corneel.cannaerts@
architectuur.sintlucas.wenk.be

Matias del Campo
Architect / Owner
SPAN
mdc@span-arch.com

Duoli Chen
Student
Aalborg University
dchen07@student.aau.de

Kenn Clausen
Student
Aalborg University
kclaus07@student.aau.de

Jeroen Coenders, PhD.
Researcher / Structural Engineer
Arup
jeroen.coenders@arup.com

Karola Dierichs
Research Assistant / Architect
University of Stuttgart
karola.dierichs@icd.uni-stuttgart.de

Milos Dimcic
Student
University of Stuttgart
m.dimcic@itke.uni-stuttgart.de

Michal Dziedziniewicz
Student
Rhode Island School of Design
mdziedzi@risd.edu

Alessio Erioli
Researcher / Architect
University of Bologna
alessio.erioli@unibo.it

Günther Filz, Dr.
Assistant Professor, Architect
University of Innsbruck
guenther.filz@uibk.ac.at

Moritz Fleischmann
Assistant Professor, Architect
University of Stuttgart
moritz.fleischmann@
icd.uni-stuttgart.de

Christoph Gengnagel, Dr.- Ing.
Professor
Architect / Structural Engineer
UdK Berlin
gengnagel@udk-berlin.de

Odysseas Georgiou
Researcher / Structural Engineer
University of Bath
odysseas.georgiou@hub.com.cy

Jörg Huber
Research Assistant
UdK-Berlin
jhuber@udk-berlin.de

Hossein Jamili
Board Member
Achivision Company
jamili@archi-vision.com

Poul Henning Kirkegaard
Professor
Aalborg University
phk@civil.dk

Axel Kilian, PhD
Assistant Professor
Princeton University
akilian@media.mit.edu

Clemens Klein
Student
TU Berlin
ctklein@hotmail.com

Jan Knippers, Dr.-Ing.
Professor / Structural Engineer
University of Stuttgart
j.knippers@itke.uni-stuttgart.de

Joy Ko
Professor
Rhode Island School of Design
jko01@risd.edu

Toni Kotnik, Dr.
Senior Researcher
EHT Zürich
kotnik@arch.ethz.ch

Jan-Peter Koppitz
Structural Engineer
Ove Arup & Partners
jan-peter.koppitz@arup.com

List of Contributors

Oliver Krieg
Student
University of Stuttgart
oliver@davidkrieg.com

Mathias K. Kristensen
Partner and Creative Director
Adapa ApS
mathias@adapa.dk

Arne Künstler
Structural Engineer
Imagine Structure GmbH
kuenstler@imagine-structure.eu

Johan Kure
Student
Aalborg University
jkure07@student.aau.de

Oliver Labs, Dr.
Research Assistant
Saarbrücken University
mail@oliverlabs.net

Guillaume Labelle
Doctoral Assistant
EPFL Lausanne
guillaume.labelle@epfl.ch

Lorenz Lachauer
Research Assistant
EHT Zürich
lachauer@arch.ethz.ch

Elisa Lafuente Hernández
Research Assistant
UdK Berlin
lafuente@udk-berlin.de

Stefan Liczkowski
Student
TU Berlin
stefan.liczkowski@googlemail.com

Daniel Lordick, Dr.
Professor
TU Dresden
daniel.lordick@tu-dresden.de

Blandini Lucio, Dr.
Senior Project Manager
Werner Sobek Stuttgart
lucio.blandini@wernersobek.com

Rupert Maleczek
Research Assistant
University of Innsbruck
rupert.maleczek@uibk.ac.at

Martin Manegold
Structural Engineer / Partner
Imagine Structure GmbH
manegold@imagine-structure.eu

Thiru Manickam
Student
Aalborg University
tmanic07@student.aau.de

Sandra Manninger
Architect / Owner
SPAN
ms@span-arch.com

Alex Matl
Architect
soma
mail@soma-architecture.com

Achim Menges
Professor
University of Stuttgart
achim.menges@icd.uni-stuttgart.de

Julien Nembrini, Dr.
Scientific Researcher
UdK Berlin
nembrini@udk-berlin.de

Christoph Nytsch-Geusen, Dr.
Professor
UdK-Berlin
nytsch@udk-berlin.de

Martin Oberascher
Architect / Partner
soma
oberascher@soma-architecture.com

Norbert Palz
Professor
UdK Berlin
n.palz@udk-berlin.de

Mark Pauly, Dr.
Professor
EPFL Lausanne
mark.pauly@epfl.ch

Thomas Pearce
Student
TU Berlin
pearcethomas@yahoo.com

Alexander Peña de Leon
PhD Student
RMIT University
alexander.penadeleon@rmit.edu.au

Gregory Quinn
Structural Engineer
Ove Arup & Partners
greg.quinn@arup.com

Armin Mottaghi Rad
Architect
Archivision Company
mottaghi@archi-vision.com

Christian Raun
Partner
Adapa ApS
christian@adapa.dk

Steffen Reichert
Research Assistant
University of Stuttgart
steffen.reichert@icd.uni-stuttgart.de

Judith Reitz
Associate Professor
RWTH Aachen University
reitz@gbl.rwth-aachen.de

Matthias Rippmann
Research Assistant
ETH Zürich
rippmann@arch.ethz.ch

Thilo Rörig, Dr.
Researcher / Discrete Geometry Group
TU Berlin
roerig@math.tu-berlin.de

Stefan Rutzinger
Architect
soma ZT GmbH
rutzinger@soma-architecture.com

Steffen Samberger
Student
TU / UdK Berlin
Steffen.samberger@gmail.com

Fabian Scheurer
Associate Researcher / Partner
designtoproduction
scheurer@designtoproduction.com

Kristina Schinegger
Architect / Partner
soma ZT GmbH
schinegger@soma-architecture.com

Volker Schmid, Dr.-Ing.
Professor
TU Berlin
volker.schmid@tu-berlin.de

Albert Schuster
Project Manager
Werner Sobek Stuttgart
stuttgart@wernersobek.com

Yuliy Schwartzburg
Doctoral Assistant
EPFL Lausanne
yuliy.schwartzburg@epfl.ch

List of Contributors

Stefan Sechelmann
Doctoral Assistant
TU Berlin
sechel@math.tu-berlin.de

Enrique Sobejano
Professor
UdK Berlin
enriquesobejano@nietosobejano.com

Werner Sobek, Dr.-Ing.
Professor
Werner Sobek Stuttgart
istuttgart@wernersobek.com

Achim Benjamin Späth
Research Assistant
University of Stuttgart
spaeth@casino.uni-stuttgart.de

Andre Sternitzke
Research Assistant
UdK Berlin
sternitzke@udk-berlin.de

Leo Stuckardt
Student
TU Berlin
leostuckhardt@googlemail.com

Martin Tamke
Associate Professor
CITA, Royal Academy of Fine Arts
Copenhagen
martin.tamke@karch.dk

Mette Ramsgaard Thomsen, PhD
Professor
CITA, Royal Academy of Fine Arts
Copenhagen
metteramsgaard.thomsen@karch.dk

Kemo Usto
Student
Aalborg University
kusto07@student.aau.de

Liss C. Werner
Associate Researcher
DIA, HU Berlin
liss@tactile-architecture.com

Andreas Woyke
Student
TU Berlin
andreaswoyke@yahoo.co.uk

Printing: Ten Brink, Meppel, The Netherlands
Binding: Stürtz, Würzburg, Germany